Microworld Simulations for Command and Control Training of Theater Logistics and Support Staffs

A Curriculum Strategy

December 1998

*John R. Bondanella, Matthew W. Lewis,
Paul S. Steinberg, George S. Park, Dina G. Levy,
Emile Ettedgui, David M. Oaks, Jerry M. Sollinger,
John D. Winkler, John M. Halliday,
Susan Way-Smith*

Arroyo Center

*Prepared for the
United States Army*

Preface

As the Army evolves into a "force projection Army," its ability to deploy quickly and conduct missions away from its garrison location places increasing importance on effective combat service support (CSS) command and control (C2). The need for training CSS management skills necessary to be effective in an increasingly information-rich and distributed environment provides the opportunity to reexamine training for support unit staffs above the division level.

The Deputy to the Commanding General of the U.S. Army Combined Arms Support Command (CASCOM) asked the Arroyo Center both to examine the current training of combat service support staffs' C2 skills and to explore what new approaches to training might be necessary for the emerging Force XXI support units.

This report discusses one new approach—a process approach to cost-effectively train these C2 skills for future logistics operations—focusing on the future training structure, content, and methods such an approach entails. This report should interest Active and Reserve Component Army logisticians in field units, as well as in the institutional training base.

The research was conducted in the Manpower and Training Program of RAND's Arroyo Center, a federally funded research and development center sponsored by the United States Army. The research was conducted, documented, and approved for public release in 1998.

Contents

Figures

Tables

Summary

Introduction

Under the auspices of Force XXI, the Army is in essence going through a process of reengineering itself and evolving into a "force projection Army," a process that stresses the ability to deploy quickly and conduct missions away from its garrison locations. Such changes place increasing importance on effective combat service support (CSS) command and control (C2). These challenges and changes to how CSS management will occur in an increasingly information-rich and distributed environment provide the opportunity to reexamine training for support staffs and determine how the Army might change its training to best prepare for new styles of CSS management.

This report argues that the current structure, content, and methods of training high-level CSS staffs will not answer the needs of the Force XXI Army and proposes an alternative approach—entailing changes in structure, content, and methods—based on a "process" view of training. Changes in methods in particular focus on the use of microworld models: small-scale simulations of organizations and operations.

Shortcomings of the Current Approach to CSS C2 Training

Based on our experience, reviewing several publications that form the basis for higher-level staff planning, interviewing staff involved in training, and attending a series of training events and a contingency to witness training in action, we find that the current approach to CSS C2 training has a number of shortcomings— shortcomings that are exacerbated by the need to meet the requirements of Force XXI. In essence, the current approach is "project oriented," focused around preparing for and executing a staff training exercise and with less emphasis paid to the content of what is learned.

However, the shortcomings are not just in training content; they are also in training structure and methods. Specifically:

- **Structure:** Exercises are not integrated during a training year or across training years;

- **Content:** Exercises are focused too narrowly, concentrating on the sustainment aspects of an operation only *after* the theater has matured and not training the skills of efficiently managing a dynamic system;

- **Methods:** Exercises, which are large and simulation-based, focus on the warfighter, not the logistician, are not intensive and realistic enough, and are costly to maintain and run to support exercises.

A Process-Oriented Approach to Army CSS Training

Given these shortcomings and our understanding of logistically oriented private-sector businesses, we argue that the Army should be focusing on a more process-oriented approach to CSS staff training. Instead of focusing on organizational relationships and segments of performance within stovepiped functions, the process approach takes an end-to-end view of a process, which enables those being trained to view the impact of their function on end-to-end performance and to understand how other segments in the process might affect their function. Such an approach entails making changes to the current structure, content, and methods of training.

Changing Training Structure

We argue that the training structure must be rethought to focus on the overall goals of the logistics units and provide more appropriate, integrated training opportunities. For example, training events early in the annual training cycle might focus on early-entry aspects of building a logistics infrastructure for an operation. Subsequent exercises might then focus on the build-up phase, sustainment management, and redeployment. The learning goals from each exercise should lead into the goals for the next one. This structure should provide for more learning to occur within existing time and resources.

Changing Training Content

The content of training, i.e., the knowledge and skills that are the target of the training exercise, should also be shifted from the current emphasis on the accounting of current assets to proactive management of assets with realistic planning horizons. This is a major shift in training emphasis, from trying to account for the current state and chase down missing materiel, to managing a dynamic system and looking ahead to minimize bottlenecks and delays. Such management skills are reportedly rarely practiced in current exercises, partially because of the limited amount of simulated days.

Changing Training Methods

The methods most appropriate for teaching the new content may vary substantially from the current methods and tools. Evidence from the commercial world suggests that when small groups of managers from across business functions within an organization meet to define and analyze business processes, the contact can be very enlightening and lead to new insights on how to improve these processes. As in the Army, managers often work only in one functional unit in an organization, so jointly developing a "process map" of an end-to-end business process can be informative for all involved.

There is also anecdotal evidence that learning the dynamics of how a complex business process can operate, complete with feedback loops and variability, can develop the management skills to deal with a wide variety of novel business situations. Private-sector organizations have used computer-based microworld models to teach managers about the business processes for which they are responsible and to help them understand how certain variables in their businesses relate to certain outcomes.

Using Microworld Models to Train Processes: Building and Pilot-Testing Prototypes

To gain a better understanding of how microworld models might be used in a staff training environment, we developed three prototype microworld models. Those prototype models simulate the National Training Center (NTC) repair parts order cycle; an early-entry module site selection and construction process for a theater support command; and a contingency operation Class IX theater distribution network. These prototype models highlight different vantage points: viewing the whole process from "above," taking the god's-eye view; looking at the process from "within," as one of the nodes inside the process; and interacting as a networked process model.

We have conducted small pilot studies with staff members from CSS operating units to demonstrate how a process approach might be implemented in a training event. By allowing us to gauge our materials, questionnaires, and training schedule, the pilot studies have provided sufficient information and confidence in the training approach to enable us to carry out a larger-scale demonstration. The results of this successful demonstration are being reported separately in another document.

Conclusions and Implications

The Army's challenge is to design a CSS training strategy that can be implemented under conditions of personnel turbulence, split-based operations, increased reliance on information, and decreased training resources faced by the Army of today, as well as that expected in Force XXI and beyond. Given this challenge and these constraints, a process approach to CSS training is appropriate and useful. In addition, smaller-scale microworld models based on commercially available software can be used within this process approach to train Force XXI CSS operational concepts and to reflect on the CSS processes necessary to implement those concepts in a contingency.

Finally, the curriculum and prototype microworld model approach proposed here has applications beyond the CSS training environment. Our prototype microworld models focus on "glass box"—as opposed to "black box"—modeling, where staff being trained can actually can see the underlying rules and change those rules as they change assumptions about the environment in which they can expect to be operating. Such versions of microworld models are appropriate to any organization that needs to train staff under distributed conditions in uncertain environments and to avoid the time- and resource-intensive costs of bringing staff together for a large game at a central location.

Acknowledgments

The authors wish to thank a number of individuals and organizations for their assistance and input to this research. Especially beneficial has been our close working relationship with the 310th Theater Army Area Command (TAACOM) and the 311th Corps Support Command.

MG Thomas Plewes, TAACOM Commander until December 1996, and his staff significantly aided our understanding of the emerging Theater Support Command doctrine, hosting our interview sessions during the Yama Sakura and Prairie Warrior 97 exercises. The 310th TAACOM staff provided a continuing exchange of ideas throughout 1996 and 1997, especially Colonel James Rowland, Deputy Chief of Staff, G3; Colonel Margaret Tankovich, Acting Deputy Chief of Staff for Support Operations; and LTC Patrick Cathcart, G3 Operations Officer.

We had significant input from MG John Crowe, Commander, 311th Corps Support Command (COSCOM), and his staff when they hosted us at two simulation-based exercises (Cascade Steel 95 and Keen Edge 96). We were invited to attend all command briefings and staff meetings. The open access to conduct individual interviews of the 311th COSCOM staff provided many insights, which we continued to build on over the past two years. LTC John Brault and LTC Glenn Mudd were most helpful in coordinating our visits.

Colonel James Paige, Joint Logistics-Advanced Concept Technology Demonstration, sponsored visits to organizations using the Logistics Anchor Desk in support of Operation Joint Endeavor (OJE). His sponsorship enabled us to interview logisticians throughout the OJE chain of command, and to visit units in Bosnia and Hungary.

MG James Wright, Commander, 21st Support Command, BG (P) Larry Lust, Deputy Chief of Staff for Logistics, U.S. Army Europe, and LTC (P) Gary Addison, G4 1st Armored Division and Task Force Eagle, shared their insights during Operation Joint Endeavor when we visited them and their staffs in January, May, and June 1996.

Colonel Charles Ennis, Deputy Commander, U.S. Army Japan/9th TAACOM provided open access to interview his staff during Yama Sakura XXXI and shared his perceptions as he mentored the exercise participants.

Having access to expertise and information on the Combat Service Support Training Simulation System has been invaluable, and we thank Mr. Larry Wilson, National Simulation Center, Fort Lee, for his continuing support. The input of logistics trainers from the National Training Center has also informed and influenced our work, particularly COL Wayne Taylor and LTC Claude Shipley and other past and present members of the Goldminer Observer/Controller team.

We are indebted to several colleagues for their input to this research. David Oaks authored the National Training Center repair parts order cycle microworld model while attending the RAND Graduate School. RAND colleagues Douglas Merrill and Jon Grossman were instrumental in the formulation stages of this research. Fran Seegull was instrumental in furthering our understanding of simulation use in coursework at Harvard Business School. We also wish to thank Dave McArthur and Marc Robbins for their detailed comments and suggestions, which have substantially improved this report.

Nikki Shacklett's editorial support and publication assistance were most helpful. Sandra Petitjean and Mary Wrazen provided us with graphics support. We thank Joyce Gray for her assistance in preparing the manuscript, Christine Hillery for assisting in the pilot study administrative preparations, and Patricia King and Joy Moini for technical assistance during the pilot studies.

The authors are solely responsible for any information and errors contained in this report.

Abbreviations

AAR	After-Action Review
AC	Active Component
APOD	Aerial Port of Debarkation
APOE	Aerial Port of Embarkation
ASL	Authorized Stockage List
AWE	Army Warfighting Experiment
BCTP	Battle Command Training Program
C2	Command and Control
CASCOM	U.S. Army Combined Arms Support Command
CBS	Corps Battle Simulation
CONUS	Continental United States
COSCOM	Corps Support Command
CSS	Combat Service Support
CSSTSS	CSS Training Simulation System
EAD	Echelon Above Division
FSB	Forward Support Battalion
FMC	Fully Mission Capable
ISB	Intermediate Support Base
JTF	Joint Task Force
LAD	Logistics Anchor Desk
LOGCAP	Logistics Civil Augmentation Program
MESL	Major Event Simulation List
METL	Mission Essential Task List
MOS	Military Occupational Specialty
MSB	Main Support Battalion

MTP	Mission Training Plan
NMC	Non-Mission Capable
NTC	National Training Center
ODS/DS	Operation Desert Shield/Desert Storm
OJE	Operation Joint Endeavor
OR	Operational Readiness
PLL	Prescribed Load List
POD	Port of Debarkation
POL	Petroleum, Oil, and Lubricants
PW	Prairie Warrior
RC	Reserve Component(s)
RSOI	Reception, Staging, Onward Movement, and Integration
SARSS-O	Standard Army Retail Supply System-Objective
SIMEX	Simulation Exercise
SIPRNET	Secret Internet Protocol Routing Network
SME	Subject-Matter Expert
SOP	Standard Operating Procedure
STAMIS	Standard Army Management Information System
STAARS	Standard Army Action Review System
TAACOM	Theater Army Area Command
TF	Task Force
TPFDD	Time Phased Force Deployment Data
TRADOC	U.S. Army Training and Doctrine Command
TSA	Theater Storage Area
TSC	Theater Support Command
TTP	Tactics, Techniques, Procedures
TY	Training Year
USAREUR	U.S. Army Europe
YS	Yama Sakura

1. Introduction

Background

During the Army's most recent large-scale contingency—Operation Desert Shield/Desert Storm (ODS/DS)—the United States deployed large numbers of people and massive amounts of equipment and supplies, and did so quite impressively. In addition to occurring in a very narrow time window, the deployment occurred in Saudi Arabia, where there was no established combat service support (CSS) infrastructure comparable to what the Army had counted on for potential deployments to Europe during the Cold War. Moreover, managing the buildup of the theater CSS infrastructure was complicated by the fact that there was no overall theater command and control (C2) support unit. The absence of such a C2 support unit was aggravated by the fact that the CSS support comprised active and reserve units, individual ready reservists, and host nation support.

During the Army's planning for its more recent, but smaller contingency operation, the deployment to Bosnia and Hungary as part of Operation Joint Endeavor (OJE), the support community faced the same type of organizational situation: the doctrinal support for such an operation would most likely consist of a corps support group. However, such a group could not handle the myriad of host-nation, joint, and combined operations and coordination needed to support the deployed combat force. The Army deployed a theater army logistics C2 element to the joint task force in Hungary to support the operation. This C2 element consisted of Active Component (AC) and Reserve Component (RC) organizations and individual ready reservists working together in an ad hoc organization; the framework was a hybrid of staff sections from the 21st Theater Army Area Command (TAACOM), part of U.S. Army Europe and 7th Army, and the 3d Corps Support Command, part of U.S. Army V Corps. Some of the reporting functions were to the 3d Corps Support Command (COSCOM) main headquarters in Wiesbaden, Germany, while others were to the 21st TAACOM in Kaiserslautern, Germany.[1]

---

[1]Based on author observations and discussions with various staffs during visits to Army and joint organizations in Germany during January 1996 and in Germany, Hungary, and Bosnia in May and June 1996.

Despite the success of such operations, some very detailed accounts of Operations Desert Shield/Desert Storm (ODS/DS) indicate that the mission in ODS/DS was accomplished much more by creativity and brute force than by closely following existing CSS doctrine and practices.[2] For example, the Army Central Command (ARCENT) created the 22nd Support Command as an organization to integrate the logistics effort in theater. That command operated much differently from what Army doctrine and practices specified at the time. For example, Army doctrine would have provided logistics support to units behind the corps boundary through a TAACOM, and other support would have been provided by selected theater army groups or other formations not a part of the TAACOM. Instead, the 22nd Support Command became involved in planning all aspects of the operation; this included supporting the theater rear area, establishing logistics bases in the corps area, and even providing theater logistics support to accompany the coalition attack into Kuwait. The support command concept was not new to the Army. The Army had used theater support command organizations in the past, but it had not included the support command in the "Army of Excellence" force structure in the 1980s. As the Army analyzed new support requirements after ODS/DS, it concluded that even the former theater army support command did not address all the situations that expected future environments might encompass, such as split-based operations and politically imposed ceilings on the quantity of people to be deployed in a given contingency.

Thus, subsequent to ODS/DS, the Army began to consider a different form of CSS management. One conceptual response—the Theater Support Command (TSC)—is being tested and developed in the Army's Force XXI program. The TSC has the ability to conduct split-based operations and to deploy command and control C2 headquarters in various modules. It has an early-entry module, somewhat larger functional modules as the theater expands, and a fully staffed functional command headquarters as required by the size and mission of the force.[3]

The TSC concept is evolving as a consequence of Force XXI experiments and analysis. The Army has selected the 310th Theater Army Area Command as the

_____

[2]These accounts come from several sources: *22d Support Command's Operation Desert Shield After Action Report*, Volume XI, Tab K ("Go to War IPR," December 16, 1990), Kingdom of Saudi Arabia: Headquarters, 22d Support Command, 1991; John J. McGrath and Michael D. Krause, *Theater Logistics and the Gulf War*, Alexandria, VA: U.S. Army Materiel Command, 1994; and Major General Martin White (ed.), *Gulf Logistics*, London: Brassey's (UK) Ltd., 1995.

[3]*Draft Concept for Support Command and Control at Echelons Above Corps* (Fort Lee, VA: U.S. Army Combined Arms Support Command, January 31, 1996), p. 3.

Force XXI experimental unit for CSS above the corps.[4] As the TSC and other Force XXI concepts mature, new training strategies need to be developed for CSS staffs at all levels. Since the TSC is an emerging concept, it has no particular training or exercise programs established. However, there are programs and exercises that address portions of the missions a TSC would be assigned. Therefore, we studied the Army approaches to training in several of the more relevant exercises over the past two years: I Corps/311th Corps Support Command's Exercises Cascade Steel (1995) and Keen Edge (1996), the 310th TAACOM portion of Prairie Warrior 96 (May 1996) and Prairie Warrior 97 (May 1997), a Reception, Staging, Onward Movement, and Integration (RSOI) Process exercise in Korea (April 1996), and U.S. Army Japan's Bilateral Exercise Yama Sakura XXXI (January 1997). We also discussed the Army's staff training approach in a series of interviews in January, May, and June 1996 with CSS units involved in Operation Joint Endeavor (OJE).

Logisticians we met with during those exercises and contingency operation pointed to significant shortcomings in the opportunities to train large CSS unit staffs. More specifically, they (and we) identified areas for improvement in the *structure* of how such training is organized and managed, the *content* of the training (i.e., in the kinds of knowledge and skills emphasized), in the *methods* used to train, and in the *resources* with which to implement a training program:

Structure: The exercises are not integrated during a training year or across training years. Further, the staffing can be, and many times is, different across the exercises in any given year. Consequently, whatever learning occurs by staff members in one exercise is not necessarily translated into learning for staff members participating in other exercises.

Content: The exercises are focused too narrowly, tending to concentrate on the sustainment aspects of an operation *after* the theater has matured and ignoring the critical training that CSS staffs need in the early stages of building a theater infrastructure. They also do not train the skills of efficiently managing a dynamic system with an understanding of end-to-end processes.

Methods: Large, simulation-based exercises generally focus on the warfighter, not the supporters, and thus do not give the CSS staffs intensive, realistic play. These large simulations are also costly to maintain and run to support exercises. Beyond these observed training shortcomings, the Army also faces the problem of not having the resources to expand staff training

---

[4]Discussions with MG Thomas Plewes, Commander, U.S. Army Theater Army Area Command, during a RAND briefing presented to him on September 6, 1996, by John Bondanella and John Winkler.

beyond current levels, because the Army is still reducing units. This lack of resources, together with personnel turbulence within the Army, is particularly acute in RC units, especially for staff training events that are outside the 15-day annual unit training.

The Army needs a new approach to CSS training that overcomes the observed shortcomings of the current approach, that addresses staff personnel turbulence, and that is effective within the current resource constrained environment.

Objectives and Scope

This report discusses a new approach to CSS C2 training to support Force XXI organizations and missions and the changes to structure, content, and methods needed to implement it. It also discusses the building and pilot-testing of prototype microworld simulations that are part of the proposed approach.

In terms of scoping issues, the focus of this work (and of this approach) has been to develop training for CSS *staffs operating as a staff, not for individual training.* Individual training issues are outside the scope of this work, although we comment on them where relevant to developing necessary staff skills.

In addition, our initial focus has been on the large unit staffs: corps headquarters and support commands as well as theater headquarters and support commands in echelons above corps. We categorized these organizations as "allocators and providers" and the other division/nondivisional units as "consumers." Although the distinction is somewhat blurred—because allocation and provision of services occurs at all echelons—we made the distinction because we observed that the higher-level staffs get very little exercise in situations involving relatively long time horizons. When staffs train they need to be focused on plans that project into an uncertain future that goes beyond a set of fixed tasks in a tightly bounded set of conditions and a short time frame. Thus, the corps and theater army organizations seem to be good focal points for long-horizon planning but still with a strong execution emphasis. Below that level, units are more focused on shorter horizons and daily execution.

Finally, the prototype microworld models described in this report are just that— prototypes. While they have been, and are being, pilot-tested, they have yet to be validated in controlled experiments; such validation is expected to be a part of future work.

Approach

In terms of developing the new approach to CSS C2 staff training, we relied on reviews of emerging doctrine, contingency and exercise after-action reports, site visits, and interviews. We reviewed Force XXI emerging doctrinal concepts that addressed how the future Army might organize and conduct military contingency operations. With these concepts as a framework, we focused on how support might be delivered and what that implied for future training. To understand the problems with current Army CSS organizations and practices, we visited training organizations and interviewed commanders and staffs; to understand current methods and approaches for staff training, we visited Army schools and interviewed staff, collected and analyzed mission training plans (MTPs), and observed exercises; to understand existing and planned simulations and models, we examined simulations and their means of delivery; and, finally, to gain a comparative perspective on how logisticians are trained in the private sector, we visited and interviewed firms, corporate training organizations, and software developers.

We then developed a prototype curriculum we believe is appropriate for future training. The curriculum depends on a new approach to training structure, content, and methods. We identified microworld models as a significant enabling method for training new concepts. In developing the prototype microworld simulations, we deliberately worked with commonly available commercial simulation software and provide three exemplar cases—the National Training Center (NTC) repair parts order cycle (developed for a previous project), an early-entry module site selection and construction model, and a contingency operation Class IX distribution network model. Specifically, we used two commercial software products: "ithink" from High Performance Systems, Inc., and "Extend" from Imagine That, Inc. These products were used for illustrative purposes to build alpha and beta prototype microworld models of several support unit processes. Our intent was to illustrate the feasibility and effectiveness of such microworld models in support unit staff training. The use of these programs in our research does not constitute an endorsement of those commercial products, nor does it indicate a preference for those over any other commercial products currently available. During subsequent research, we will develop criteria that might be useful in selecting commercial products with which to construct support unit microworld models for training.

Finally, we conducted pilot studies to refine the microworld models. Individuals and small groups of CSS subject-matter experts (SMEs) participated in the pilot studies, which addressed TSC doctrine and operations and were based on both process-mapping exercises and experiences with the dynamic microworld models.

Organization of This Document

Section 2 briefly discusses the shortcomings of the current Army approach to CSS C2 training, with a focus on structure, content, and methods. Section 3 discusses a proposed curriculum strategy and what changes it entails for the structure, content, and methods of training. Section 4 discusses the development of microworld models as a key enabler of the curriculum strategy and our pilot studies of them. And Section 5 presents some conclusions and extrapolates on the usefulness of the approach in other Army organizations and in the private sector.

2. Shortcomings of the Current Approach to CSS C2 Staff Training

In this section we briefly examine some of the problems our research has uncovered with the Army's current approach to training CSS C2 staff.

Current Approach to Army CSS C2 Staff Training Is Project-Oriented

As discussed in Section 1, in trying to understand how the current approach to Army CSS training operates, we reviewed several publications that form the basis for higher-level staff planning, interviewed a number of the staff involved in training, and attended a series of training events and a contingency to witness training in action. Based on that experience, we found that the current approach to CSS training is "project oriented," focused around preparing for and executing the exercise. Less emphasis is given to the content of what is learned. More specifically, the training focuses on organizational relationships and segments of performance within stovepiped functions; the training stresses how each staff section performs its own function, with only limited coordination with one element above it or below it in the hierarchy. As a result, the training audience never gets to view its impact on end-to-end performance of all segments of a process nor how other segments in the process might affect the training audience. Staffs are generally trained in crisis response and in solving individual problems, with the result that the approach tends to be reactive: Staff train to react to a specific event in a specific scenario.

Such an approach creates a number of problems, problems that can be discussed in terms of the structure, content, methods of training, and resources. In Section 1, we summarized the problems in these areas; here, we briefly expand on that discussion.

Structural Problems: Exercises Are Not Integrated During Training Year or Across Training Years

Issues surrounding training structure include the ways the training events were—or were not—linked across the course of the training year. Some events appeared to be tied to the needs of the warfighter training exercises (e.g., Yama

Sakura and Prairie Warrior). Another (Cascade Steel) directly aimed at exercising a set of support staffs within a small window of support operations (three days of play in a mature theater support environment).

Figure 2.1 is a timeline that shows an example of this lack of integration. In many units, members of the staff do not all attend the same exercise: The figure represents a typical situation where some members attend Prairie Warrior, while others attend Yama Sakura or a Reserve Component Simulation Exercise. The planning, coordination, and train-up for each of these exercises is conducted by and for the designated attendees of each exercise. Consequently, the staff as a whole does not always get trained on staff functions across the spectrum of their mission, only on a more narrowly focused set of tasks and skills specific to the exercise for which they are preparing.

Although unit staff members we interviewed said that valuable training takes place in such training events, they also recognized that the opportunity to learn is not as great as it could be. This stems from the fact that the events are not linked to each other or to a larger "curriculum" structure. The skills learned do not then build on one another over the course of a year and cover the spectrum of operations that would be encountered across the course of an actual deployment. This approach to training is an opportunity to perform skills assumed to be already learned; in fact, the participating staff may not all be proficient in these skills because of personnel turbulence and assignment to duties in an exercise that they have not performed before. The "train-up" period in Figure 2.1 is oriented to learning the mechanics of the exercise and computer support aspects of a simulation, not to staff missions and functions.

Part of the reason for this is the apparent lack of "ownership" of the overall training process during the course of the year. The unit commander has the responsibility for defining the training goals for exercises, but the commander is constrained in several important ways. First, large support unit participation in exercises is often tied to the warfighter exercises where support units get limited play; what play they do get is limited to manipulating data and building realistic briefing slides to present to the warfighters. All reports point to staffs not getting much opportunity to solve the kinds of problems they would regularly encounter in planning and establishing support in actual operations.

Second, not all commanders are expert at assessing skills, strengths, and weaknesses and at designing appropriate training experiences to meet those requirements. Such assessment and design can be complex and is not part of their training. CASCOM and TRADOC provide simulation and training resources, but the design of training experiences during the year is not a strength

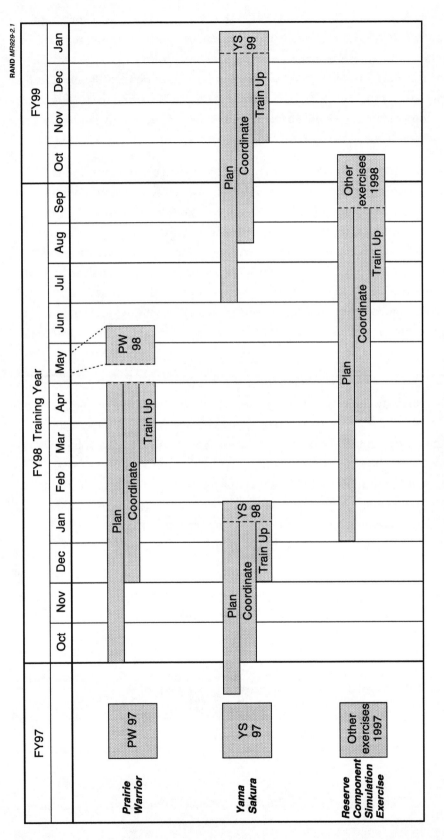

Figure 2.1—Training Event Timeline

10

in the current system. There are events in the exercises from the Major Event Simulation List (MESL) that can be tailored to meet specific training requests or requirements, but these are small-scale, local events as opposed to pieces of a long-term training strategy and structure that go across the year. Further, in the RC especially, participation in the various exercises is by different subsets of an organization's staff: The whole organization does not necessarily get to play in each exercise. Thus, whatever learning does occur is uneven across the staff.

Not only are the training experiences not integrated across the year, but the staffing of training events themselves is constrained in some important ways. First, the exercises do not provide much training for the time invested. Figure 2.2 shows the amount of training time provided by one simulation—CSS Training Simulation System (CSSTSS), which (as we shall discuss) provides a moderately realistic set of data for CSS staffs at division and corps levels to analyze and manipulate during exercises. Apart from the quality of this simulation, a 15-day exercise does not provide 15 days of training in staff functional performance. The figure illustrates roughly how the training hours are spent during a 15-day training period.

Note that most of the training time is spent on learning how to use the simulation in the scenario and training environment and on administrative preparation and travel. Further, during the five or so days when the simulation is used, the training may be limited to day-shift operations (primarily because of the lack of resources to conduct 24-hour operations). So for the five days of the exercise proper, only 60 hours at most involves staffs performing their functional duties. Of that 60 hours, there is no systematic documentation of how many hours are deemed by the training audience staffs to be relevant to the duties they will be conducting during an actual operation.

Second, there is the issue of who is actually available to participate in exercises. When we look at what percent of a unit is trained and how many hours are spent per month on learning CSS management skills, as opposed to other tasks and

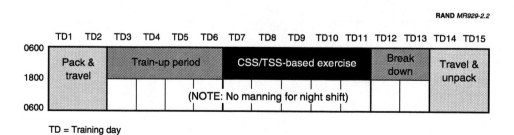

TD = Training day

Figure 2.2—CSSTSS Exercise Length and Training Time

requirements, we discover two things. The first is that there are few hours of opportunities to practice the core staff skills in the professional management of an Echelon Above Division (EAD) unit. The second is that the entire unit never exercises together for any significant period of time. Because of exercise schedules, funding constraints, and the availability of individual staff members (e.g., conflicts with either personal, nonreserve activities or with other reserve commitments, such as participation in ongoing contingency operations), the entire unit does not participate in all exercises. This problem is compounded by high turnover rates—a problem encountered throughout the entire Army.

The RC units also raised issues about the appropriateness of their prior training and military occupational specialty (MOS) to the tasks they were expected to perform in the exercises.

Having discussed issues involving the structure of training, we now turn to the content of what is trained during current exercises.

Content: Exercises Are Focused Too Narrowly

The current simulation-based training exercises provide a number of indirect benefits, including learning how the overall organization is structured, who the sister units are, and getting to know and trust the other units involved in the exercise—benefits that are highly valued by unit commanders. In addition, as mentioned earlier, good training opportunities occur with MESL events, but they are few, strongly pre-scripted, and very limited in scope.

However, observations and interviews also consistently identify shortcomings in the content of what skills are taught and emphasized in current training. The skills that appear to be most often emphasized are described in mission essential task lists (METLs) with the terms "manage," "monitor," "protect," and "sustain." However, the amount of time spent on such activities is small compared with the total time committed to the exercise.

Several other types of skills appearing in unit METLs are not emphasized in current training. These include "design," "analyze," "negotiate," "coordinate," "promulgate policies," "provide instructions," and "monitor execution." There are three reasons for this: (1) The scenario is always a mature theater, so there is little design and policy communication. (2) The time window of simulated play is limited. (3) The training largely emphasizes reporting what is current versus proactively trying to predict demands or bottlenecks; this is somewhat driven by the use of simulators which emulate the Standard Army Management Information Systems (STAMIS) and by the lack of good automated data synthesis and analysis tools.

12

Beyond the limitations in the skills that are taught, the exercises themselves neglect the dynamic complexity of missions in teaching skills: their breadth, depth, and time horizons. This is illustrated in Figure 2.3. A large CSS staff's planning horizon is weeks to months, while its daily operational focus is on events that might affect operations in the next 72–96 hours. Clearly, training events with a time horizon of 60 hours—such as the CSSTSS-supported exercise discussed above—do not exercise the staffs adequately. Furthermore, the 60 hours is normally focused on sustainment operations, not on the other aspects of a CSS staff's mission and responsibilities. These other aspects include the very important functions of analyzing the warfighter's operational campaign plan to determine the risk of alternative logistics courses of action, setting up theater infrastructure, designing theater distribution systems for materiel, establishing CSS services policies, and allocating support units by phases of an operational plan. Thus, the phenomenon of "dynamic complexity" is not portrayed: The exercises do not contain decisions that would have been made by the staffs of the playing organization, nor those in higher or lower echelons, during earlier phases of the contingency.

Thus, the staffs do not learn how to design an infrastructure and support operation; instead, they end up reacting to crisis events and making ad hoc decisions. To the extent that the staff needs crisis management training, this is

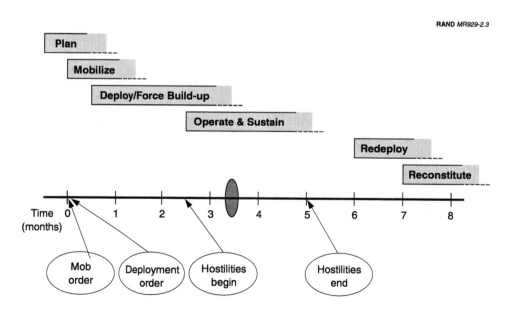

NOTE: The shaded oval represents the extent of a typical operation played during a training session.

Figure 2.3—Exercises Neglect Dynamic Complexity

necessary experience; however, it is not sufficient for training the staff in the breadth of duties and missions that it would encounter during the course of a contingency operation.

The content of current training exercises also has important limits in the number of activities being managed and directed in parallel. Most exercises focus on the combat aspects of a contingency. To train combat skills, this may be a proper focus; it certainly may even be a proper focus for small CSS staffs at division, brigade, and battalion headquarters.

However, large unit CSS staffs at the regional or theater level must focus simultaneously on establishing and sustaining the combat operation as well as supporting the other major aspects of joint task force operations. Some units may be arriving during the course of combat operations, and the TSC operations staffs must focus on RSOI at the same time they focus on the combat operation. Simultaneously, the TSC planning staffs need to be oriented on later phases, when the combat operation changes in intensity, when forces disengage, and when forces redeploy from the region.

The new skills that need to be taught—skills that emphasize dynamic planning and longer time horizons—demand the use of new, appropriate methods.

Methods: Large, Simulation-Based Exercises Not Focused on CSS Staffs, Not Realistic, and Too Expensive

The methods for teaching the knowledge and skills needed by CSS managers at EAD are largely independent of the content itself. That is, the "how" is not necessarily related to the "what." For example, Civil War–era CSS doctrine and operations could be taught with computer simulation or ODS-era CSS doctrine and operations could be taught with vugraphs and lectures.

One of the primary methods for teaching CSS management skills and knowledge, apart from having individuals read doctrinal literature, is through the use of large-scale, simulation-based exercises. Based on visits to several of these training events and extensive interviews with the members of staffs (the training audience), the simulation teams, and Simulation Center staff members, there appear to be a number of limitations to the current simulation-based exercises.

First, the exercises vary widely in the focus and frequency of the training provided to support units. Many exercises include the support units only as entities that provide information to bolster the play of warfighter units; fewer exercises focus training on skills and missions of the support units themselves.

The 311th Corps Support Command exercise Cascade Steel, which used the CSSTSS simulation discussed elsewhere, was the exercise that received the most praise from the staffs involved in the training. CSSTSS was developed to simulate STAMIS input and output and provide large quantities of data for the support unit staffs to analyze and manipulate. CSSTSS does not provide a forward-looking component nor give the staffs an opportunity to design processes during the course of training. The processes exercised are either designed as built-in parts of the simulation or are developed by a neutral control staff.

Additionally, the entire staff does not get trained from these exercises. For example, Cascade Steel is held every other year. It uses the CSSTSS simulation, which the staff believes is a great training tool. However, only a small segment of the staff benefits from this simulation more frequently than once every two years; that segment comprises staff members who are able to play in the annual Combined Arms Center Prairie Warrior Exercise on a recurring basis. This situation is even more complicated with the TSC, because the TSC early-entry module and functional modules are composed of staff from other functional commands. After the exercise, there is very little opportunity to come together again as a staff to discuss and reflect on their learning experiences, and to participate in training that reinforces or increases their skills.

The Army experiences considerable turnover in all positions and organizations. Consequently, the head of a staff section may have little familiarity with the spectrum of influence that section may have on the entire process. In several exercises, we have observed that the staff members may either be newly assigned or simply "filling a slot" temporarily because of the unavailability of the regular staff member.[1]

A second limitation is that the simulations used in the exercises do not give CSS staffs at echelons above division (EAD) realistic, intensive play. Part of the limitation is caused by the emulation of the CSS STAMISs, and part is caused by the resultant information flows. The focus of existing CSS exercises is on simulated reports from the STAMIS emulation. Because the STAMISs are oriented toward providing post-hoc data about inventories and trends in services provided, the exercises tend to focus on reports of accounting-like information

---

[1]We have spoken with a variety of staff members at different exercises and even in actual deployments. Sometimes, having staff who are somewhat familiar with the function is preferable to not having a staff member at all to fill certain positions. While this is a learning situation for some people, little learning takes place when the staff member is identified at the last moment and is not able to prepare to operate in the position. This is not an intentional effort to limit the quality of the exercise, but rather reflects the situation where there are more demands on the unit for particular skill levels than there are fully qualified staff members to fill the positions.

about materiel and on aggregating information to pass to higher echelons. The other part of the limitation stems from the interfaces into the simulation-based training tools themselves. They do not lend themselves to a focus on proactive planning for CSS aspects of contingencies. Current simulations are not only data-intensive, they also take considerable time to develop and load the data for a specific scenario (we have observed up to six months of advance coordination and data preparation). Consequently, the current simulation-based exercises do not challenge players to develop multiple plans and contingencies, the databases are inadequate for generating challenging scenarios and event lists, and the use of data-exchange technology is not adequate to effectively support visualization of the logistics processes being played.

Further, the automated tools (e.g., radio frequency tags and other in-transit visibility enablers; standard army retail supply system-objective (SARSS-O); joint total asset visibility program) postulated as enablers for the effectiveness of the TSC are not simulated in those exercises. Although there are demonstrations of these tools available during the exercise, they are not well integrated into the play. Players learn what might be available, but they really do not learn the power such tools have to assist in the TSC management functions. We did note staffs using various features of the developmental Logistics Anchor Desk (LAD), but these were limited compared to the powerful applications the LAD could offer.[2] In the TSC exercises, LAD was used primarily as a research tool to determine quantity and location of supplies in stockpiles. In our visits to OJE units, we observed LAD being used on a broader basis to support Time Phased Force Deployment Data (TPFDD) analysis and to locate critical supplies.

Beyond the lack of realism, the large, simulation-based exercises also do not provide enough intensive practice time. As shown in the earlier discussion on the CSSTSS simulation, the amount of training time spent on actual CSS operations management relative to the entire time spent on the task is small— only 5 out of the 15 days, or about 60 hours.

Third, because of scarce training hours, preparation time, and other resources, such large-scale simulations are very expensive to use. One indicator of this expense is that a higher-level logistics unit might only be able to participate in an exercise driven by a large-scale simulation every two years. Taking into

---

[2]The Logistics Anchor Desk was an Advanced Concept Technology Demonstration (ACTD) program conducted by the Army under the joint auspices of the Under Secretary of Defense for Acquisition and Technology and the Assistant Secretary of Defense for Logistics. The LAD suite of tools included software that provides access to a variety of databases and models, such as the radio frequency tags database, Knowledge-Based Logistics Planning System (KBLPS), the Analysis for Mobility Platform (a suite of strategic and operational level transportation models), time-phased force deployment data (TPFDD), and Army Total Asset Visibility databases.

consideration the high personnel turbulence experienced across the Army (approximately 30 percent per year in a unit, and even higher for individual positions within a unit), this lengthy time between simulation exercises does not provide continuity of training for the staff.

Resources: Force XXI Staff Training Curriculum Strategy Must Consider Resource Constraints

Our approach to developing a curriculum strategy assumes that there will be no significant increase in training resources currently available and that the curriculum would be implemented within the skill sets generally available to the Army. The Army had not published training resource levels for Force XXI at the time of our research. However, our approach to structure, content, and method indicates that the curriculum could most likely be implemented by changing focus rather than by significantly increasing resources over current levels.

3. A Process-Oriented Approach to Army CSS Training

As the brief review in Section 2 showed, there is a need for a new approach to training military CSS staff. Here, we discuss a proposed new approach—more process oriented and focused on having individuals understand how their particular CSS functions fit within the overall CSS process. After establishing the need for this sort of approach, we examine how firms in the commercial sector— where process-oriented training is common—train their employees, and we discuss the kinds of changes this approach implies for Army CSS training structure, content, and methods.

The Need for a New Process-Based Approach to Training CSS

Under the auspices of Force XXI, the Army is in essence going through a process of reengineering itself. Such reengineering emphasizes the rapid availability of information, much of it obtained in a nonhierarchical flow. For example, units involved in Operation Joint Endeavor have already begun to use new technology and procedures to improve the flow of information. Operational reports have been put on a web page for viewing by anybody within the chain of command, from the originating unit to the Department of the Army. This has been done on the Secret Internet Protocol Routing Network (SIPRNET), a classified system, and requires appropriate access approval. The information generated by radio frequency tags is also available on a web site. Commanders, both warfighters and CSS commanders, know this technology is available and are now asking for more predictive information rather than inventory/accountability information. We noted during the conduct of Yama Sakura XXXI that the acting TSC commander mentored the player staff both to look out over a longer time horizon and to use the availability of more timely and detailed information to influence the course of events.

As Force XXI units deploy in a more modular fashion, these information technologies are intended to enable smaller staffs to not only keep abreast of the situation but also use information to help shape their support operations. Current deployments and Force XXI concepts stress a smaller "footprint" during the initial stages of a contingency; consequently, staffs who deploy early in the contingency—especially CSS staffs—are expected to have a broader understanding of the total unit mission. While they cannot be functional experts in every area, they will be expected to know the points of necessary coordination. In some cases, the early-deploying staffs will be performing some portion of C2 tasks normally performed in a different operational unit; these staffs may not be familiar with either the unit or the tasks. Thus, the Force XXI training curriculum would need to consider C2 staff training that is focused on a better understanding of the entire enterprise, rather than on individual functional training. Consequently, we focused our research on approaches and models that would help staffs design and manage processes rather than learn how to execute one particular process. Further, the processes designed for a contingency may have some common characteristics but usually would not be stable, steady-state activities like those typical of a commercial business; the operation of the process would involve much more uncertainty than a commercial firm would encounter in operating its business processes.[1]

The challenges and changes to CSS management brought on by Force XXI reengineering provide the opportunity to reexamine how training for EAD support staffs should be done. How might the Army change its training to best prepare for these new styles of CSS management?

Lessons Learned from Private-Sector Logistics Practices

One of the likely sources for new approaches to CSS training is what commercial logistics firms—where the pace of innovation has remained quite high over the past 10 years—have done as they have gone through a reengineering process similar to the one the Army is undertaking now. We reviewed business process reengineering literature and then visited several private-sector firms to enhance our understanding of how firms were training their employees about their new business enterprises. While some information was gleaned from how businesses approach training structure and content, we found very little pertinent to staff

---

[1]When we say that commercial firms do not encounter uncertainty, we are not referring to uncertainty of the market, but rather to uncertainty about how a particular process ought to be operated. The military has doctrine, which acts as a guideline under many circumstances; however, because of the uncertainty of the operational environment, processes in a contingency may not follow that doctrine closely, especially in the early stages of a deployment.

training exercises and more about individual training methods. Consequently, our reviews and interviews—and thus our findings—were oriented primarily to gaining an understanding of training methods in two areas:

- What simulation models are businesses using to train their personnel in the business enterprise?
- Given the use of those simulation models in training, do businesses have quantitative performance measures to show improvements in training effectiveness?

How Businesses Are Using Simulation Models to Train Their Staffs in the Business Enterprise

We found that businesses generally use models to train their staffs how to perform a particular function or process. These models are oriented toward specific training rather than toward learning systems-dynamics, process thinking. This type of training in the business community focuses on an individual's learning about the firm's organization, functions, and reports. There is some movement toward using models that portray a process approach to the business, and these models simulate processes that cross organizational and functional boundaries. We characterize these as "black-box" models—they contain rules and processes designed by some central group, and the training audience is supposed to learn how to operate within the process.

Some commercial firms use multimedia approaches, centered on computer-based training, to help employees think through their role in specific processes. These approaches, which center either on influence diagrams or interactive microworld model simulations of only one aspect of a business activity, do not seem to address the systems-dynamics aspects of end-to-end processes. Again, while they use simulation models, these are black-box models where the trainee gets to react to different situations presented during the simulation of a process.[2]

At a more advanced level, the trainee would be able to make some changes in the process as it is simulated (e.g., "turning the dials" to control a simulated flow). These type models are akin to "management flight simulators," as discussed by Peter Senge et al. in the book *The Fifth Discipline Fieldbook,* one of several guides to understanding and redesigning organizations. In fact, the book argues that the real benefit of these "management flight simulator" business models comes from

---

[2]Brandon Hall, *Return-on-Investment and Multimedia Training,* a research study funded by Macromedia, Inc., San Francisco, CA, 1995.

the activity of defining the process. The authors call this activity "understanding the business" and point out that there is much institutional learning to be accomplished in this step alone.[3]

A different type of learning occurs when the training audience can use a "glass-box" model—a model in which the training audience can see what the rules, measures, and processes are and can change them to fit a variety of situations it might need to explore.

Several microworld simulation model tools[4] have been developed that make the glass-box model approach feasible for training staffs. With the confluence of object-oriented simulations, enhanced graphic-user interfaces, and significant increases in computer hardware speed and capacity, this microworld model provides a new dimension to training.

However, our observations indicate that the business world has not yet adopted these tools widely for analytic applications; there is even less use for training applications. To determine why these tools are not in greater demand, especially for training, we pursued this topic in a literature review and in discussions with several commercial firms that provide process model tools. The only place where we found any quantitative information was in a 1996 *CIO Magazine* article.[5] That article included a projection done by the Gartner Group, Inc., based on a survey of organizations using business processing reengineering tools. The survey itself was not available to us, but the *CIO Magazine* report of it stated that

> one in four of those companies are using or thinking about using products with advanced simulation or animation capabilities. But as the higher-end tools come down in price and lower-end tools improve, those numbers of users should rise. According to the Gartner Group report, by 1998 more than half of all BPR [business process reengineering] projects will make use of some form of simulation or animation.

The *CIO Magazine* article also reports that

---

[3]Peter M. Senge, Charlotte Roberts, Richard B. Ross, Bryan J. Smith, Art Kleiner, *The Fifth Discipline Fieldbook: Strategies and Tools for Building a Learning Organization*, New York: Doubleday, 1994.

[4]From the companies that provide commercially available modeling tools for desktop/personal computer applications, we obtained information on several applications, either from product purchases for our own use, Web sites, company literature, or magazine articles. The applications we gained the most familiarity with are Extend, by Imagine That! Inc., and ithink, by High Performance Systems, Inc.; we achieved this familiarity by purchasing the authoring version of their software and building prototype logistics process models. Many of the Web sites provide demonstration versions of their product and thumbnail sketches of customer "success stories." We have listed in the references the Web address of these success stories from Imagine That! Inc.; Powersim; Gensym Corporation; Simulation Dynamics, Inc; and Sandbox Inc.

[5]Daniel Gross, "Endless Possibilities—Tools That Simulate 'What if' Scenarios Show Managers That a Picture Is Worth a Thousand Words," *CIO Magazine*, February 1, 1996, extracted from the publication's World Wide Web site *(http://www.cio.com)* on August 11, 1998.

[f]or the near future, most users will fall into three categories: internal consultants, charged with documenting their company's organizational processes; external consultants who create new models to optimize processes for clients; and internal and external consultants who are specifically modeling work flow applications. Regardless of who is using them, simulation tools should become more widespread as hardware and software costs drop and the technology improves. As companies ramp up, they face a bewildering array of choices. About 30 vendors offer products ranging from $600 to $16,000, and some customized versions of these tools can run up to $60,000. In a market that has no truly dominant players, the capabilities of these simulation tools vary greatly. . . . The highly sophisticated technology behind Vega was until recently considered out of reach to all but the most well-heeled customers for creating flight simulators in the aerospace and military industries. "The technology that used to cost tens of millions of dollars can now be delivered to users' desktops for $30,000 to $50,000. That has caused us to rethink how it can be applied in other areas," says Smoot [Mike Smoot, vice president for sales and marketing at Paradigm Simulation].

We followed up on this topic in further Web site research (the site addresses are listed in the bibliography) and then in telephone conversations with the two companies doing simulation modeling that seemed to us the most pertinent to Army training needs. In the Web site research of companies that advertise process model simulation and consulting, the applications cited were primarily for specific business analysis, not for more generalized training. In the telephone discussions and in our literature/Web-based research, we posed two questions: Were these companies engaged in any staff training development, for either private or government organizations? Why were companies, in their opinion, not using microworld models for staff training?

Our conversations indicated that the business world has not used these simulations in wider applications for a variety of reasons, expense being one. (This was also reported in the *CIO Magazine* article.) In some cases, the simulations are very data-intensive and very expensive. For example, Final Bell uses actual stock market databases. In other cases, there has not been much thought given to applications beyond "factory assembly line" processes and sequential tasks within a specific process rather than across processes;[6] model

---

[6]We have observed this in a variety of Web-based articles of software customer applications. These applications tend to focus on very specific, local processes. One very interesting exception is the experience of CEMEX, a Mexican cement company, which used Gensym Corporation's G2 software to integrate operations management of cement mixing and delivery in Guadalajara; the application integrated aspects of customer demand and delivery times, production, and transportation on a daily basis (Peter Katel, "Bordering on Chaos," *Wired Magazine*, July 1997, as reported on the Gensym Web site *(http://www.gensym.com)*, August 15, 1998). A more typical view of single-focus process applications can be seen in the other "success stories" on Gensym's web site. The size of models and databases can be relatively unwieldy in practice if one is looking to replicate processes in high resolution. For example, an Extend simulation of a space launch process became

developers and consultants have developed significant business in those applications and have not ventured into the training domain. Software companies have not advertised the staff training aspects; consequently, business customers have not seen the potential and thus have not demanded such applications as tools to train their personnel as staff entities as opposed to individual training. However, we did observe that these tools are starting to be used in academic institutions for individual training. High Performance Systems, Inc. has listed several examples of their model applications to support case studies produced for Harvard Business School Publications Company.[7]

How Businesses Measure Improvements in Training Effectiveness

We have found some quantitative data about how businesses measure more traditional approaches to multimedia or computer-based training. For example, one report shows how several major firms have reduced training time in traditional classrooms compared to multimedia approaches on the order of 60 percent and have reduced total training costs on the order of 40 percent.[8] Data exist that describe how computer-based training and multimedia approaches have significantly increased effectiveness and reduced the resources for training individuals using black-box models.[9] However, we have not found a body of literature that describes in quantitative terms how glass-box models have improved training in the context of an overall curriculum. We have found qualitative statements about how such microworld models help in analyzing decisions.[10]

Two other salient aspects from the private-sector literature review helped us shape the Army CSS C2 staff training curriculum: the focus on end-to-end

---

too large for a single application and was subsequently used in smaller, stand-alone sections (author telephone conversation with Larry Krause, Simulation Dynamics, Inc., on August 11, 1998).

[7]High Performance Systems, Inc.'s primary Web page is at *http://www.hps-inc.com.* Selecting "About HPS" leads to "Learning Environment Products." Selecting one of the products, e.g., "Building Service, Driving Profits: RGP Financial Services," leads to "Harvard Business School Publishing Company" *(http://www.hbsp.com/frames/groups/cases/new/new_products.html).*

[8]Hall, op. cit., p. 2.

[9]There are several studies cited on the Society for Organizational Learning's Web site *(http://learning.mit.edu/res/SoLres/AN3_698.html)* and through Dr. Ernst Diehl's biographic page *(http://learning.mit.edu/com/peo/ediehl.html).* Most of the entries discuss research on how organizations learn or describe "learning laboratories," but they present no statistical data on the use of simulations in staff training exercises.

[10]See Gross, op. cit. Also see the "customer success stories" on various software providers' Web sites. PROMODEL Corporation's Web site *(http://www.processmodel.com/customer/stories.html)* describes the use of its product ProcessModel by Eli Lilly, Prudential, and Newport News Shipbuilding. Powersim's Web site includes customer success stories at Ford, Nexus, and British Telecom *(http://www.powersim.com/html/f_success_ford.htm, _nexus.htm,* and *_employ.htm).* Computer Aided Process Improvement's Web site *(http://www.capi.net)* describes its work with the United States Postal Service using the dynamic modeling and simulation tool Extend (from Imagine That! Inc., *http://www.imaginethatinc.com).* All these sites were active and accessed on December 10, 1998.

processes, and how those processes are measured. Many firms use a paradigm of supplier/process/customer and apply it to many *internal* processes. These firms focus on supply-chain management and emphasize end-to-end processes (e.g., customer order to receipt) rather than a traditional functional or organizational perspective (e.g., how does the sales group perform, how does the warehouse group perform, how does the shipping department perform) that we have seen in the Army.[11]

The process approach entails significant differences in how performance is measured. Unlike the Army's more functionally driven, project-oriented approach, where performance is measured on how well a specific subsegment of the process performs, the private sector increasingly is focusing on activity measures or process measures that assess sets of related activities. This focus on activity measures moves the organization away from functional optimization and toward a broader, more integrated management system that looks at the upstream and downstream effects of key activities. In addition, the measures are not static. Instead, they are continually changed to reflect the concept of continuous improvement. The measures also evolve based on changes in the firm's environment.

If the Army is to translate to a more process-oriented approach to CSS management, it will have to make changes to training structure, content, and methods.

A Case for the Army's Use of Simulations to Teach Logistics Management Skills: Addressing Training Needs Not Found in the Corporate Sector

This report has cited evidence of two specific ways that corporations currently use simulations: to support teams making strategic planning decisions, and to train individuals. There is also anecdotal evidence that although the market for training staffs in management skills will be increasing, that market is not yet robust. Why, then, should the Army lead the commercial world in embracing a new set of methods and technology tools to teach logistics management skills to individuals and staffs?

Since the 1940s, the military has consistently led industry in piloting new methods for training, including the application of new technologies (Gray, Pliske,

---

[11]This situation in the Army is changing significantly as the Army's Velocity Management program is maturing. However, these changes have not found their way into CSS C2 staff training and exercises.

and Psotka, 1985). A recent study for the White House's Office of Science and Technology Policy suggests that DoD research spending on learning technologies continues to dominate similar spending by other non-DoD federal sources such as the National Science Foundation and the Department of Education (Glennan, Bodilly, Lewis, McArthur, and Moini, 1998). The military has been willing to invest heavily in research in learning technologies because it has urgent needs that differ in important ways from the needs of commercial firms. These needs include the reality that in peacetime the military has no natural access to practicing its "core competencies" or "business processes" on a daily basis. In contrast, most businesses operate daily and practice is continual as part of normal work. For some organizations, like EAD logistics units, there is no set of organization or infrastructure to practice with, since logistics infrastructures are rapidly designed and built to support a specific contingency when a mission arises. There can be no "on-the-job" training when the job does not exist most of the time.

Another key factor that drives the military's interest in more effective training, supported by technology, is the high turnover rate of personnel in jobs. Since officers generally spend two years or less in a job, and enlisted personnel on average also have short periods of service, there is a very strong need to quickly train skills and then get people into their jobs and productive with as little delay as possible. There is little time to amortize the initial training costs. On the commercial side, our discussions with commercial aircraft maintenance organizations and with a commercial mining operation revealed the importance of investing large amounts of resources and time in employee training, both formal and on the job. In the aircraft industry, new mechanics fresh out of school are not expected to be paying for themselves for four years or more as they gain experience and skills. The military has strong pressures to train quickly and efficiently, since turnover is high.

Finally, there is an especially strong need among military reservists to be able to train teams to work together during a contingency when in peacetime they are geographically distributed. Having tools that allow group interaction and team practice via networks provides training opportunities that are rare in normal reserve training experiences.

The issue of the high costs of data-intense models is not relevant for the kinds of microworlds we support for this training. Large amounts of actual data can be approximated well by mathematically derived distributions of simulated data, and the models themselves are not large. The argument is that you do not need exact data or very fine-grained models to learn how to generally manage complex, dynamic systems.

These three needs cited above—the lack of on-the-job training opportunities, the need for rapid, effective training, and the need to train teams to collaborate when they are geographically distributed—argue for the military to again lead the way in the art and science of training. Although the simulation-based training market is growing in industry, the military has greater and more immediate needs for better methods to quickly and effectively train management skills. One way to address those needs is through application of microworld-based curricula running on networks to support management skills training.

Changing Training Structure

When changing to a "process" perspective, what emerges is a different, more integrated approach to designing the sequential set of learning experience. This new view emphasizes continuity in staff development across multiple years of a training schedule.

In the current training strategy, exercises and learning are not linked across the entire staff, as pointed out in the earlier section on training shortcomings (see Figure 2.1). The various exercises are attended by different participants, training is limited to the sustainment phase of an operation, and other skills required to design and implement a theater CSS plan are not exercised. Figure 3.1 expands on the current structure portrayed in Figure 2.1 to show how a unit might design a "linked learning view" of training over three years. (This same structural construct will be used again later in Figure 3.4 to illustrate changes in method of training.) During training year 1997 (TY97), opportunities existed for the TSC (Provisional) to participate in three major Army exercises: Yama Sakura XXXI (YS 97) with U.S. Army Japan, Prairie Warrior 1997 (PW 97) with the Combined Arms Center, and Reserve Component Simulation Exercise 1997 (RC SIMEX 97).

Two key features of the TSC are its ability to command and control support units during a contingency on a modular basis and to rely on modern information technology for enhanced effectiveness. However, each of the TY97 exercises focused only on the sustainment phase of a contingency. Thus, the TSC staff did not get to learn much about how it would perform in two large portions of its joint task force support mission in a contingency: establishing a theater/regional infrastructure with an early-entry module, and operating that infrastructure with scaled-down functional modules for C2.

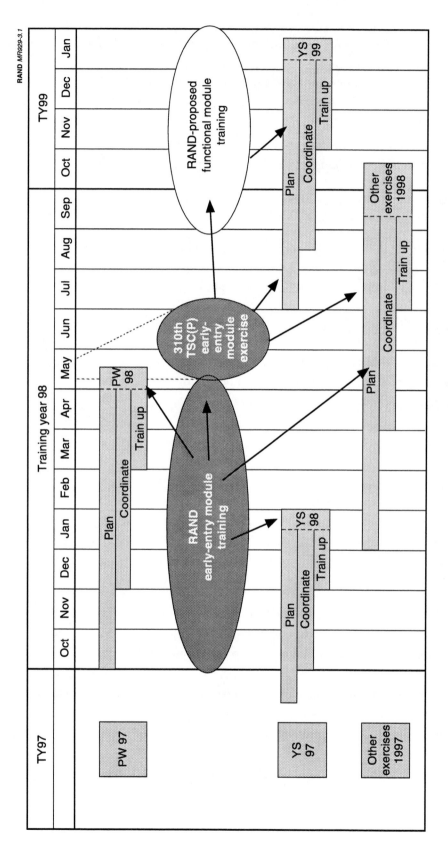

Figure 3.1—A "Linked-Learning View" of Training Events

The process view of training would focus on how the early-entry and functional modules could be exercised during the training year. We believe this can be accomplished with a process approach to training management and training content. To illustrate this approach, RAND Arroyo Center conducted pilot training in TY97 and TY98 with the 310th TSC (Provisional). The first ellipse in Figure 3.1, "RAND early-entry module training," illustrates our concept that staff functions and missions (e.g., "early-entry module") need to be trained independent of a particular exercise. The RAND project team conducted several sessions from June 1997 through April 1998 with a focus on the distribution management responsibilities of the TSC during the early-entry module phase of an operation. The second ellipse in Figure 3.1, "310th TSC(P) early-entry module exercise," shows that the 310th TSC then conducted a two-week exercise for its staff to study and to draft procedures that the TSC might use during the early-entry phase of an actual contingency. The procedures were not tied to a specific scenario, but rather to the processes that the early-entry module might be expected to perform, regardless of scenario. These processes then could be modified to meet the specific requirements of any given scenario.

The intent of this approach is then to have the TY97 and TY98 events be used as a prototype for the 310th TSC(P) in formulating its training and exercises for TY99. The third ellipse in Figure 3.1, "RAND-proposed functional module training," portrays the continuation of this training approach into TY99.

Changing the structure of how training is designed not only enables the integration of different training events, as pictured in Figure 3.1, but also allows a unit to train both the specific staff members participating in the exercises and those unable to attend. The training could be focused on the breadth of missions (early-entry, expanded theater, etc.) and tailored to meet the entire staff's training needs, not just the needs of any particular exercise. Staffs that work together can assess their specific strengths and weaknesses and feed these training needs to the training designers and developers. This "customer-driven" approach to tailoring instruction can only occur if the staffs have been able to function together long enough to understand their consistent behaviors and to identify skill deficiencies.

Changing Training Content

A change in training content involves more than just changing scenarios and MESLs for exercises; it requires *defining the process* and *identifying the skills and knowledge* needed to carry it out. In terms of defining the process, training should

28

address support processes, not single events. Processes are continuous and have a defined set of inputs and outputs.

Defining the processes themselves is a learning experience for staff. Defining the process to be trained is neither simple nor quick. It often begins with understanding the doctrinal literature surrounding the process. But since processes often involve multiple units or cut across traditional Army doctrinal boundaries (e.g., across transportation and supply in the case of distribution processes), this can involve some detailed syntheses of information from multiple sources. Process definition also includes iterations of interviews with the stakeholders in the process, that is, with the people who have actually tried to carry out the process and can articulate the standard operating procedures (SOP), tactics, techniques, and procedures (TTP), and "common sense" that underlie the implementation of the high-level description in the doctrinal literature. If possible, the detailed description should not only capture the flow of materiel and information, but also the C2 flows and the flow of other enabling resources (e.g., personnel time and finances). A graphical diagram or "map" of the process is extremely useful in making explicit many of what might be implicit assumptions about the process.

The second step involves identifying the needed management skills. As the process is being defined, the skills necessary to manage it will begin to emerge. Some of the skills are individual ones that may be better learned in individual training. However, as discussed in Section 1, our focus here is on group as opposed to individual training, so we concentrate on staff skills: What should this staff group be expected to do and how well? The expected performance of the process needs to be identified. The ultimate performance measurement should be oriented on the output of the process, not on any particular segment. However, the segments also have to be measured to identify where the process needs to be improved, something we discuss later in the section on changing training methods.

Below we illustrate this process definition and skill identification using a simple example of a process within an Army operational mission. The TSC, as the highest-level support organization in theater, works closely with the Army Component Commander of a Joint Task Force to develop theater policies and practices during the planning stages and to monitor the execution of those policies and practices during the deployment and contingency operations. At the start of an operation, forces deploying to a theater of operation go through the "reception, staging, onward movement, and integration" (RSOI) process.

Process Definition: Understanding the RSOI Process of Developing a Theater Infrastructure

From a top-level perspective (as shown in Figure 3.2), the RSOI process of developing a theater infrastructure has a defined set of inputs (guidance, resources, and information) and a defined output (units ready for operational missions). The important elements for the TSC headquarters (the support headquarters) are to design the subprocesses, to provide resources for those subprocesses, and to monitor the execution by operational support units. The TSC can then measure its performance by comparing its outputs with the requirements specified by the Joint Task Force Commander. As the TSC monitors the process—e.g., comparing what the commander asked for with what he received—it should examine whether any particular aspects of the process are liable to cause a problem, or whether the entire process needs to be reengineered.

In the context of a CSS process, we are concentrating on how the RSOI process for developing a theater infrastructure might be trained in an exercise with a microworld model, which we discuss in detail in Section 4. To keep the example simple, we develop a map of one subprocess that is critical for the TSC early-entry module to design: select and establish storage sites for supporting the theater development.

Figure 3.3 shows an example theater storage area (TSA) ammunition site design and construction RSOI subprocess; the gray-shaded parts signify the "selecting and establishing storage sites" subprocess. The TSC both designs the process

RAND *MR929-3.2*

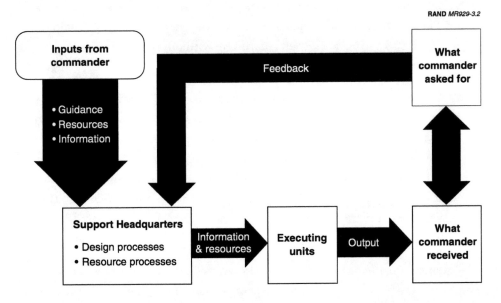

Figure 3.2—Define the Process to Be Trained: Top View of RSOI Process

30

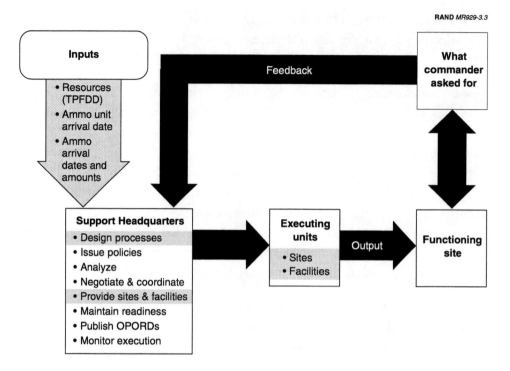

Figure 3.3—TSA Ammo Site Design and Construction RSOI Subprocess

and monitors the execution of the site establishment. The process involves several activities. It starts with an analysis of the time-phased force deployment data (TPFDD) to determine when the ammunition supply unit arrives in theater and when the actual shipments of ammunition arrive. This analysis continues to determine how much ammunition needs to be stored and issued from the site so that it can determine the appropriate size and the location relative to main supply routes. The TSC must then coordinate with other organizations, including the host nation.

The subprocess involves a variety of organizations and different staff sections within organizations. Our example highlights the main organizations involved, rather than exhaustively listing all activities in the process. It is intended to provide a perspective of how a staff might design a process that reflects a flow through time and among organizations. The "synchronization matrix" made popular in ODS/DS is a good starting point, but it is not a process map. The process map we illustrate has been used as the framework for building a prototype microworld simulation model, which is discussed in Section 4.

The example detailed subprocess map shown in Figure 3.4 was developed to emphasize the flow of information in the process, not the flow of commodities or units. The detailed subprocess map concentrates on the connectivity between

RAND *MR929-3.4*

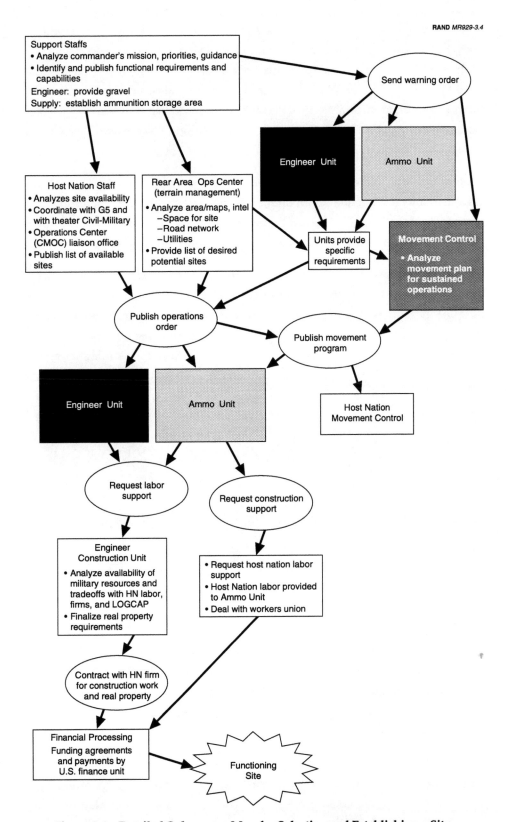

Figure 3.4—Detailed Subprocess Map for Selecting and Establishing a Site

disparate organizations. The simulation model focuses on how much time it takes the staff within each organization to complete its functions and the quality of the information passed on to the next organization. It also makes obvious the need for information that is available from prearranged agreements, such as host nation support, or information that is more doctrinal, such as plans for the size of a storage area, given a certain amount of ammunition to be received, stored, and issued.

In some cases, as shown in Figure 3.4, activities flow in parallel, while in other cases they move sequentially. In both cases, we are interested in a learning environment that enables the staff to understand how the actions of each process segment affect their own activities and how the entire process functions.

Once an operations order and a movement program have been published, the process can be mapped through to its goal, establishing a functioning site. The ultimate measure is whether a functioning site is available at the specified time and is capable of handling a specified workload. As the staff understands this process, it can begin to look for ways to improve it. Also, the simulation model runs in a matter of minutes in the stand-alone version, so the staff can train on different times or different phases of the operation during a relatively short training session.

Once the staff has identified the process to be trained, it can then focus on the skills necessary to operate it.

Skill Identification: Understanding the Skills Needed for the RSOI Process

Process design and process execution require different emphasis on staff skills. The training environment needs to provide a framework in which to understand and improve these skills. Based on our review of several mission training plans concerning higher-level CSS organizations, we developed a list of skills a command and control staff may need to perform in executing the unit's mission essential task list (METL). These skills are listed in Table 3.1. We then illustrate in the next several tables (3.2 through 3.5) how these skills are emphasized from a variety of aspects (reactive versus proactive exercises; single-phase versus multiple-phase exercises; training content; training methods). Then, Table 3.6 illustrates methods that would be most applicable to training these skills.

As we discussed earlier, most current exercises focus on the TSC operating in a sustainment phase of a combat operation. The exercises are supported by large-scale, data-driven simulations. In such an exercise, the CSS system players are

Table 3.1

Basic Skills Deriving from Unit METL

Analyze
Design
Negotiate
Coordinate
Promulgate policies
Provide instructions
Protect
Monitor execution

operating primarily in a reactive mode in an almost steady-state environment with high resolution, limited time horizons, and a single phase of an operation. A process-based exercise would be characterized by small-scale simulations, low resolution, broad time horizons, and multiple phases of an operation. Table 3.2 illustrates the emphasis those sorts of exercise place on the METL skills compared to what a process-based exercise would emphasize.

The skills necessary in that phase fall heavily into the category of monitoring the situation and handling crises. However, when the TSC early-entry module is building up the theater infrastructure, a much different set of skills is needed. Table 3.3 identifies sample tasks and the necessary skills for establishing the infrastructure.

Table 3.2

How Training Exercise Approach Emphasizes Skills

Skills Based on METL	Reactive, Data-Driven	Proactive, Process-Based
Analyze	×	X
Design		X
Negotiate		×
Coordinate	X	X
Promulgate policies		X
Provide instructions	×	X
Protect	×	
Monitor execution	X	X

× = some emphasis; X = major emphasis.

Table 3.3

Sample Tasks and Skills for Establishing the Infrastructure

Skills Based on METL	Build/Expand an Operation	Execute Normal Operations	Handle Exceptions and Crises
Analyze	X	x	x
Design	X		x
Negotiate	X		
Coordinate	X	X	X
Promulgate policies	X		
Provide instructions	X	x	X
Protect	x	x	X
Monitor	X	X	X

x = some emphasis; X = major emphasis.

Table 3.4

Relating Tasks and Skills for Building Up Theater Infrastructure

Sample General Tasks	Key Staff Actions	Specific Process Management Tasks and METL Skills
Execute normal operations	Managing and maintaining	**Manage personnel flow into and within theater:** Analyze, coordinate, provide instructions, monitor, protect
Build/expand an operation	Planning, allocating resources, and executing the operation	**Establish Class V (ammunition) theater storage area (Class V TSA):** Analyze, design, negotiate, coordinate, promulgate policies, provide instructions, monitor execution
Handle exceptions and crises	Real-time planning/ decisionmaking with limited data	**Recover from chemical strike on Class V TSA:** Analyze, design, coordinate, provide instructions, monitor, protect

In addition to building up and expanding the operation, the command and control system must simultaneously execute normal operations and handle crises as they occur. Table 3.4 illustrates how tasks and skills are related across these parallel activities, with some specific process-based examples.

Note that some of the skills listed in the "management skills" column are underlined. These are the skills that appear to be necessary to accomplish the task but are either missing or not emphasized in the unit METL that we reviewed. We also did not see many opportunities to practice such skills in our observations of large, simulation-based exercises.

These same skills are necessary for the phase of a contingency where forces and resources are being withdrawn for redeployment to their peacetime location or to another contingency area. We observed during our visit to OJE that the Task Force Eagle staff dealt with local political leaders when plans were being considered to move one combat service support headquarters from Bosnia to Croatia. The staff met with local leaders to explain why the move was necessary and to prepare them for the impact on the local labor situation. The task force and its primary base operations contractor, under the Logistics Civil Augmentation Program (LOGCAP), had hired local civilians to perform many duties. In an area that had been devastated by the earlier hostilities, having a large group of people become unemployed carried a possibly significant impact. This would affect not only local political leaders, but also any humanitarian relief agencies providing services in the area.

As mentioned above, we have listed a number of skills that appear in unit METLs in doctrinal literature and some that do not, but we argue that the additional skills should be included explicitly in the unit METL, since they are critical to the unit's ability to execute its mission.

The broad spectrum of knowledge and skills that staffs must possess to be effective are represented in three categories, as shown in Table 3.5. The knowledge and skills vary in complexity across the three categories, and each category builds on the preceding ones, as in classic curriculum design skill hierarchies.

Table 3.5

Three Categories of Knowledge and Skills for Support Staffs

		Basic/Enabling	System Understanding	Managing Dynamic Systems
Content		• Terms • Organizational structure: players • Basic tasks of each organization • Point-of-contact information • Basic analysis skills • Technology use • Collaboration skills	• Connectivity of organizations • Overall structure of basic process • Flows of resources and information	• Interactions of variables • Effects of variable changes on system performance • Which variable have most leverage on system performance • How to "herd behavior" of complex process • Predictions of future trends

The different types of knowledge and skills highlighted in each category are meant to be exemplary, not exhaustive. The ultimate goal of training for EAD support units is to produce effective management skills in a staff that allow its members to influence, or "herd," the behavior of the complex process they are assigned to manage. However, staffs should also be able to predict future behaviors to some degree of accuracy over a limited time horizon.

To carry out such tasks, the staff has to be proficient in a number of basic/enabling skills listed in the first column. These range from the very basic knowledge of terminology and the identity of other cooperating units, to the skills of being able to use spreadsheets and perform simple analyses on data sets.

Different types of methods and instructional tools might most effectively train this broad array of knowledge and skill types. These methods are addressed below.

Changing Training Methods

Once the process has been defined and the skills identified, the next set of decisions surrounds the choice of instructional tools to best train those skills, which entails providing the appropriate tools, measuring the appropriate performance in training events, and providing appropriate feedback to enable improvement.

Providing the Appropriate Methods for Training Skills

Finding the right fit of training tool with skill is often the task of identifying what tool is most appropriate. Table 3.6 expands Table 3.5 to show the potential tools for each set of skills.

Currently, for example, certain basic/enabling and system understanding skills—such as identifying objects or knowing what units are in an organization—might best be taught with traditional classroom or self-study methods, while other skills—such as managing a dynamic system—might best be taught with large-scale exercises. The future methods could be additions to the current training methods or, in some cases, possible substitutions for them. Specifically, computer-assisted instruction and small-group process training methods may be useful for training basic/enabling skills. The use of microworlds is of interest to provide training in instances where a large, simulation-based exercise may not be appropriate.

Table 3.6

Appropriate Methods for Categories of Knowledge and Skills for Support Staffs

	Basic/Enabling	System Understanding	Managing Dynamic Systems
Content	• Terms • Organizational structure: players • Basic tasks of each organization • Point of Contact information • Basic analysis skills • Technology use • Collaboration skills	• Connectivity of organizations • Overall structure of basic process • Flows of resources and information	• Interactions of variables • Effects of variable changes on system performance • Which variables have most leverage on system performance • How to "herd behavior" of complex process • Predictions of future trends
Methods			
Current	• Self-study • Doctrinal literature	• Self-study • Doctrinal literature • Large exercises	• Large exercises
Future Additions	• Traditional computer-assisted instruction	• Small-group process construction of maps • Traditional computer-assisted instruction • Microworlds use	• Microworlds use

Figure 3.5 depicts the structure of the training year as described in Figure 3.1 above. It overlays on that structure the different types of tools that might be used to support the training. Instead of trying to leverage existing exercises conducted by the Army and joint warfighting communities to focus on combat arms training, it might be useful to consider other exercises that provide more focused training directly appropriate for CSS units.

The figure shows where applications of seminar games led by humans, exercises driven by large-scale simulation models (e.g., CSSTSS or the developmental WARSIM 2000), and small simulations driven by various microworlds built to emphasize specific staff skill training might train certain types of skills more effectively than the current large, simulation-based exercises alone can do.[12] We had an opportunity to conduct some preliminary pilot studies of this approach during the 1998 training year; those pilot studies are discussed in Section 4.

---

[12]If these microworld simulations are to interact with other models, the Army would require them to comply with its high-level architecture (HLA) policies.

RAND *MR929-3.5*

FYXX training year

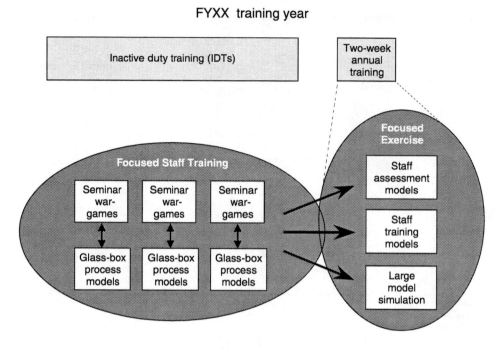

Figure 3.5—Structure of Training Year with Training Methods Overlaid

Measuring the Appropriate Performance in Training Events

Given the appropriate methods for training events, the next step is to measure performance using those methods in a training event. Measurement is critical to any method of training because it forms the basis of feedback to both learners and training designers, in turn enabling each to improve their performance. We are interested in answering the following kinds of questions: How well has the process been executed or managed by the staff? Has the process achieved the result of satisfying the commander's requirements within the specified time? If not, what are the consequences? Have the managers learned to diagnose future problems early and take appropriate ameliorating actions?

Figure 3.6 presents some hypothetical measures of the process of designing, siting, and building a TSA ammunition site, described earlier. The process of establishing a TSA ammunition site can be measured directly in how long it takes to become functional. If the schedule is not met, the consequences can be measured in terms of effect on other operations.

For example, there is a "danger zone" of not having a site to receive ammunition as it arrives at a port (air or sea). If there is too much ammunition in the port, an explosion might shut down port activities for an indeterminate time. In Scenario

RAND *MR929-3.6*

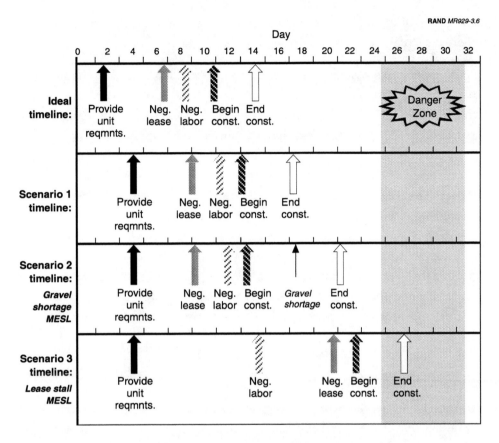

Figure 3.6—Hypothetical Measures of Process of Designing, Siting, and Building a TSA Ammunition Site

1, the timeline specified is met, and there is plenty of time to avoid the danger zone. In Scenario 2, the construction is slowed because there is a shortage of a necessary construction material, which happened because the responsible agency did not monitor the situation and ensure that gravel would be available to construct roads and ammunition pads. The effect is that the timeline moves closer to the danger zone. If any other delays occur, the project might not meet the specified goal. In Scenario 3, the contract to lease a particular site does not get approved until late, delaying the start of construction. The staff does not know how long the approval will take. Eventually, the lease is let and construction begins. However, the situation causes a delay in moving ammunition from the port; as ammunition builds up at the port, there is an explosive safety risk. This danger is caused by storing too much ammunition in close proximity without appropriate berms to guard against multiple explosions if one particular area explodes or catches fire. Thus, a set of integrated measures that looks at the total ammunition situation—arrivals, storage at the port, and storage site construction—would identify and perhaps help preclude a situation where the storage activity is entering the "danger zone."

If performance is measured only in terms of construction time, then this "danger zone" of ammunition build-up at the port is not evident to staffs dealing only with construction. However, if the entire process is measured, the "danger zone" becomes evident across staff functions of construction, supply, and transportation. In Scenario 3, ammunition stocks gradually build up at the aerial port of debarkation (APOD) and at two temporary storage sites. By day 28, stocks have exceeded the level considered safe by the ammunition safety community at the APOD and are rapidly reaching the safety level at the two temporary storage sites. Further, the temporary storage sites have a problem with security—requiring more resources to protect them than one larger site with appropriate safeguards would need. In addition, the road network servicing the two temporary storage sites is not sufficient to accommodate the projected volume of traffic over the next month as the intensity of operations increases.

Such a scenario provides a good training opportunity for the TSC Distribution Management Center. Here is a case where a microworld could provide practice not only in carrying out the process under different conditions, but also in measuring the integration of planning, analyzing, coordinating, and executing operations that cross organizational and functional boundaries, as well as the ultimate effects of not meeting the timeline.

As Figure 3.6 shows, small-scale training events should be structured so they can be conducted rapidly, allowing several scenarios to be examined. The staff should measure its improvement between cases. One of the frequent complaints we heard during our interviews was that staffs do not get to exercise frequently enough nor in a continuous environment, so they do not get a good indication of how they improved over prior training exercises. Section 4 discusses prototype models we developed. Those prototype models demonstrate how training can be effectively conducted and measured with a variety of multiple scenarios over a longer simulated time horizon.

Providing Appropriate Feedback to Enable Improvement

When to provide feedback and what to say in that feedback are universal issues in all education and training. The traditional concept of going on an exercise and then getting feedback in a formal after-action review (AAR) at the end of five days is a useful way to document major findings in the Standard Army Action Review System (STAARS) so that future staff personnel and other organizations might learn which activities should be improved or which activities are more effective and should be adopted by other organizations.

However, while the current use of such traditional AARs with the current combination of exercises and simulations might lead a staff to think about how a process might be changed for the next exercise, the situation does not provide a robust environment for staff learning, because by the time the next exercise occurs, most of the current staff may no longer be in the same positions, and the lessons become lost to time. Using a process approach to training alleviates a part of this problem but in itself is not effective enough.

Mentors would help staff sections acquire a broader perspective. The mentor should be sufficiently well versed in the process to be able to distinguish between deficient staff performance and an ineffective process. The mentor would then recommend that the staff skills be improved, or that the appropriate hierarchy change the process to overcome its weak points. In the Scenario 3 example discussed above, the mentor might be able to point out whether the contracting section is insufficiently staffed to process contracts rapidly, whether the contracting section is not exploring alternative lease sites in parallel rather than sequentially, or whether some person in the contracting office is just not able to adequately negotiate contracts.

Beyond the mentor approach, a preferred method for training might be to use the microworld exercises discussed above and in more detail in the next section—because they can be repeated rapidly, allowing a variety of cases and processes to be examined during a single training event—and to follow the AAR paradigm for tactical units at the National Training Center (NTC). There, a "Continuous Action Review" occurs during the course of an operation via observer/controllers interacting with their Blue Force counterparts. There are also AARs at every level immediately after each battle, which are sometimes referred to as "Jeep-top AARs."

4. Using Microworld Models to Train Processes: Pilot Studies of Prototype Training Curriculum

In the previous section, we discussed how microworld models—small-scale simulations of organizations and operations—might provide a useful method to train CSS processes. Here, following a brief discussion of the value of microworld models, we discuss three prototype microworld models to train CSS processes and then discuss our efforts to conduct pilot studies using those models.

The Value of Microworld Models

The Army currently uses large-scale models in its major exercises, such as CSSTSS and the Corps Battle Simulation (CBS). While they provide an abundance of data such as what would normally be produced by the Army's collection of STAMISs, such models are somewhat rigid for training purposes since they take a long time to load data and to modify processes. In addition, the players using these large-scale models are not always aware of the underlying structure of the process being modeled or of the decision rules deep inside the model. Although some learning takes place, it is limited to the same information provided by the STAMISs—accounting and inventory data.

Using microworlds for training has a number of useful attributes relative to other larger models for training CSS management skills. Such models provide an opportunity to postulate changes and then to rapidly simulate the modified or new processes to understand how they might affect the operation. In the past, this class of models has been oriented toward manufacturing and assembly-line processes; however, as more businesses are oriented on implementing new information technology as part of their reengineering agenda, current commercial microworld applications have been developed to deal with information processes.[1]

---

[1]There are a variety of microworld commercial applications, but not all of them are suitable for training large staffs. The set of applications termed "influence models" help understand the links and nodes but usually provide no way to measure activity on the links or at the nodes. Measurement is critical to learning and improving processes. Therefore, the applications that provide strong quantitative functions as well as influence diagrams are preferred. There are a variety of "success stories" published by companies that provide consulting services or process model software. We have listed in the bibliography several of these success stories as posted on a particular organization's World Wide Web site.

As such, the process approach simulated in microworld models fits well with the primary Force XXI strategy of using information technology to enhance force effectiveness by enabling the staff to understand where information technology can significantly increase effectiveness. More specifically, the staff should be able to experiment with processes to learn how the segments interact. This should not be done to "game the system"—i.e., to achieve some score—but rather to understand how behavior at one level affects another. For example, the support structure for a Joint Task Force (JTF) may not be able to handle the initial volume of sustainment supplies being shipped into the deployment area. The strategic airlift provider may not have visibility over the situation in theater and might schedule flights so that the air flow keeps supplies from stacking up at the air port of embarkation (APOE) in the continental United States (CONUS). The consequence of that action would be to send even more supplies to the deployed area when the JTF could not even handle the slower flow. If end-to-end process information were available, perhaps the strategic air transporter would decide to let the supplies stack up in CONUS or even in an enroute location, rather than sending them to the point in the deployment area that is least capable of handling the increased flow.

The 21st TAACOM (Forward) at the Intermediate Support Base and Task Force Eagle faced such a situation early in OJE. After this situation became apparent, it implemented a system to stop the large quantity of military air cargo flow enroute from CONUS through Germany and Hungary to Bosnia. That decision was based on several factors, primarily the support structure's capability and the cost of air transport into Bosnia. As military air shipments arrived in Germany, they were unloaded from the airplane. A decision was made by a 21st TAACOM representative at the airfield about which items would continue rapidly to Bosnia by air and which would go by slower ground transportation. The decision to send certain supplies by ground was made even though air was originally designated as the shipping mode.

The 21st TAACOM was hampered at the outset because it did not have end-to-end visibility of the air flow during the initial stages of OJE. Although the Defense Department and the Department of the Army had provided to the command a variety of advanced technology hardware and software intended to enable such visibility, those systems had not been fully developed nor integrated well enough by the research and development community to achieve the visibility required by senior CSS planners.

Development of Three Prototype Microworld Models for CSS Staff Training

To gain a better understanding of how microworld models might be used in a staff training environment, we used an existing RAND prototype model and developed two others: (1) the NTC repair parts order cycle model (existing); (2) the early-entry module site selection and construction model; and (3) the contingency operation Class IX distribution network model. These models highlight different vantage points: viewing the whole process from "above"— the "god's-eye view"; looking at the process from "within," as one of the nodes inside the process; and interacting as a networked process model.

The NTC Repair Parts Order Cycle Microworld Model Prototype

The NTC repair parts order cycle microworld has a number of important features. This microworld is actually a simple model that was built for two very specific purposes: (1) to enable members of the participating support units to see the overall performance of the process they were involved in (they never get to see the overall system and watch the process "perform"); and (2) to highlight the relationship between the timing of STAMIS batch cycles and the combat power of the supported Task Force. It was designed to put the learner in the "above" or "top-down" perspective of the system.

The microworld was produced quickly and inexpensively. It was designed and built with a commercial programming software system by a student in RAND's public policy graduate school. It was meant to be simple to use. As we see below, it is also a "glass box" versus "black box" simulation: The learner can see the inner workings of the glass box model and, thus, easily question and change many of the underlying assumptions.

Figure 4.1 is a screen image from the NTC repair parts order cycle microworld that shows the model itself: This is the "code" that is constructed to make the model. It is programmed graphically through a direct manipulation interface. Resources flow through links via valves and conveyors and collect in reservoirs. The model represents an older version of the automated supply control system that was in place at the time, before the new system (SARSS-O) was fielded.

The Division Main Support Battalion (MSB) and below in-theater repair parts order cycle is represented in this simulation. The "reservoir" in the upper left-hand corner (the box labeled M1s FMC) represents the number of Fully Mission

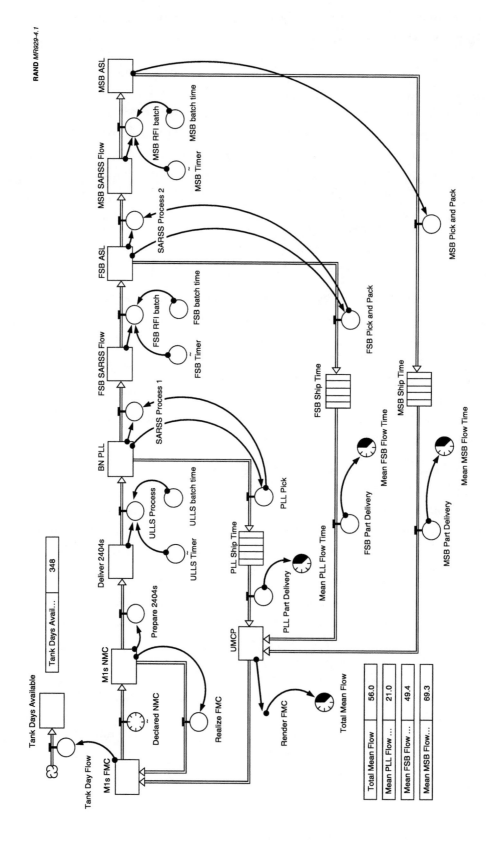

Figure 4.1—Microworld Model Screen Image of NTC Repair Parts Order Cycle

RAND MR929-4.1

Capable (FMC) M1 tanks available to the Task Force commander at any point during a rotation. The counter at the top (just above the reservoir) keeps a running total of the number of M1 days of operation realized up to any point in a rotation.

As vehicles break down, they move from the FMC reservoir into the Non-Mission Capable (NMC) reservoir and are diagnosed; parts are identified. This process takes a certain amount of time. The part requests then flow back to the echelon of supply where the part is in stock, either as part of the Prescribed Load List (PLL) at the company level, at the Authorized Stockage List (ASL) of supplies at the Forward Support Battalions (FSBs), or MSBs. The movement of information between STAMISs is critical to the speedy movement of parts to repair down vehicles. How the echelons have scheduled their STAMIS batch cycles affects how many parts are delivered and, hence, how many vehicles are repaired.

The flow of resources through this simulation is graphical: As M1s break down, the reservoir level decreases and parts graphically "flow" through the reservoirs. The staff can see the effects of batch time decisions in how smoothly the system flows. Set the batch times inefficiently and the requests move and stop repeatedly, with the result of a low number of available "tank-days" (i.e., the total number of tanks available each day across a rotation). Set them efficiently, and the system flows smoothly, with orders and parts cascading rapidly through the echelons and repair functions, resulting in more tanks repaired and in the battle the next day.

As a final note, once a process is understood, creating a simple but effective simulation to teach about it can take place fairly quickly.

The interface to the microworld is easy to use. As shown in Figure 4.2, sliders (just below the graph) are used to adjust the hour that the batch systems are run at each echelon.

The outcome graph in Figure 4.3 shows time across the bottom, the spikes depict when batches are released, and the top shows the simulated operational readiness (OR) rate changes as vehicles break, as parts begin to flow, and as repairs are achieved. Uncoordinated batches add wait times to the process by slowing part orders and delivery, which in turn slows repairs and lowers OR rates over time. This graph closely resembles the actual OR rate changes during a rotation at the NTC in 1996.

By carrying out a number of simulation runs, staff members can explore the best set of settings for a number of different scenarios that contain different variables, such as distance between echelons or speeds of travel on connecting supply routes.

48

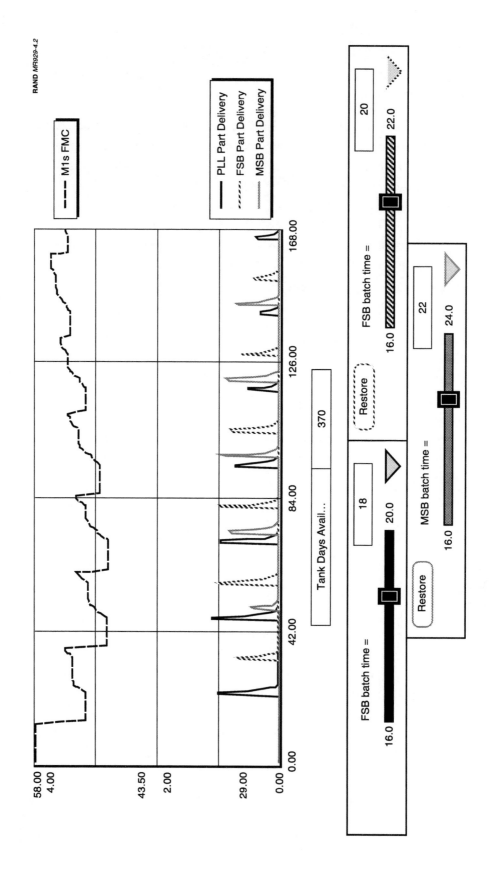

Figure 4.2—Using the Microworld Model with "Sliders"

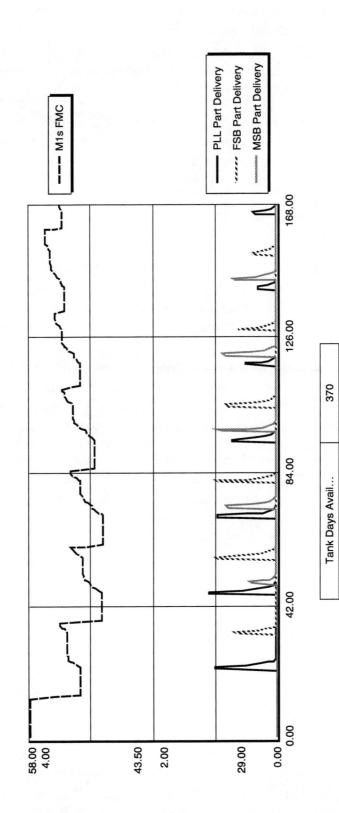

Figure 4.3—Outcome Graph from NTC Repair Parts Order Cycle Microworld Simulation

Finally, staff can go back to the dynamic simulation diagram, shown in Figure 4.1, and experiment with other aspects of the system. For example, they might try running three or more batch cycles per day or speeding up the time it takes to deliver the part to investigate the effects of changing the way the process is usually set up to perform.

These types of interactions with the internals of the models demonstrate the usefulness of having "glass box" versus "black box" simulations: The staff can see and manipulate the underlying assumptions for themselves.

Early-Entry Module Site Selection and Construction Microworld Model Prototype

A second prototype software program—based on the emerging doctrine for the TSC—has been developed that simulates the process of a TSC carrying out the mission of selecting and establishing a storage site. The microworld has two primary goals: (1) to allow staffs of units involved in the TSC to understand the process of how they would collaborate to achieve a specific mission; and (2) to run the process and see how delays and allocations of resources (e.g., personnel) could change the performance of the system. It was designed to link the performance of the process directly to an important outcome: the dangerous build-up of ammunition at temporary storage areas, described earlier in Section 3.

This prototype program was developed using commercial software similar to that used for the NTC repair parts order cycle simulation. The advantage of using such software is that the program can be made relatively simple, clear, easy to run and modify; it requires little preparation time to use; and it is available and useful at the soldier level. Of course, developing superficially simple and intuitive programs often requires complex coding at a lower level; however, the programming software can be designed so that this complexity is masked from the user, who can then concentrate on the training at hand. In fact, adequately sophisticated and available software with this level of flexibility and power is relatively new; not surprisingly, these simulation models represent the current advances of computer-based training efforts.

The components of this microworld model include elements (in the form of "agents") for Host Nation Staff, Rear Area Operations, Movement Control, Engineering Units, and Ammunition Units, each of which has distinct and necessary roles in this operation. Information (e.g., site locations, ports of debarkation (PODs), and resources available) and command instructions flow among these elements in this microworld. In the process of this operation, a storage site at a particular camp is identified, an appropriate POD is determined, equipment and labor is sent to the camp to construct the site, and (once the camp is constructed) the supplies are transported from the POD to the newly constructed site. Key parameters (e.g., type and number of supplies, availability of camp locations) are determined randomly at the outset.

Figure 4.4 shows the screen display for the program. The major portion of this display maps out the structure through which the process of selecting and establishing a storage site is undertaken. The individual boxes (nodes) with the agents inside represent the various units where orders are received and processed, work is done, and new orders go out. The lines between the nodes represent flows of information between staffs or commands. The process begins with the Support Staff box in the upper left-hand corner, which sends a Warning Order for an operation to the Host Nation Staff, Rear Area Operations, Engineering Unit, Ammunition Unit, and Movement Control Unit (the upper five boxes with agents inside). From that point, the chain of command follows the network down the process map. Note that in some cases, a unit may receive or send out multiple orders.

During the operation of this process, the program displays signals to signify progress. When a unit receives an order, an agent's clipboard inside that unit turns red; correspondingly, when the order has been fully processed, the clipboard goes back to normal. Progress is also signaled by the two graphics items in the lower right. The top map displays the eastern portion of "Lantica," a simulated area of operations for Prairie Warrior 97. When the order to send the supplies to a particular POD is received (after processing from Host Nation Movement Control), that POD is highlighted to show the arrival of the supplies. The lower map displays the activities at eight different camps in the region. When the order is sent to deliver the necessary equipment (forklifts and bulldozers) and labor to construct the storage site at a camp, that camp is highlighted with this information. After an appropriate time (which may include random delays), the site is expected to be constructed, so that the highlight for the equipment and labor turns off. Meanwhile, the highlight for the supply turns off at the POD and turns on at the camp, representing the movement of the supply from the POD to the newly constructed site.

This particular simulation can be run in either above mode, where any planner or executor can play; within mode, where principal staff play; or networked mode, where the organization network is played at a distance. Below, we discuss each type of run.

Above mode. In this mode, the user can simply run the simulation and observe the process as orders are received, processed, and passed on from an agent in one

RAND *MR929-4.4*

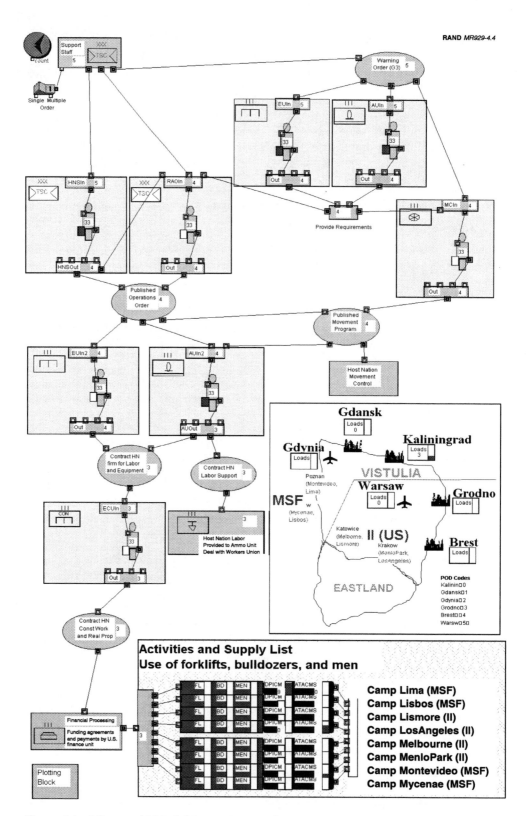

Figure 4.4—Microworld Model Screen Image of Early-Entry Module Site Selection and Construction

unit (e.g., Ammunition Unit) to the next (e.g., Movement Control). In the computer simulation, these activities are represented by animation of the agents. In fact, once a computer agent is activated, it actually opens, processes, and writes readable text files that represent orders as they are read, acted upon, and generated. Thus, either during the simulation or afterwards, the user is able to examine the orders received by the various units as well as the ensuing orders sent.

Thus, the above mode could augment training received by a planner or operator. While observing the process in operation, the user can track flows and identify bottlenecks. The lessons learned could be made more sophisticated by running multiple orders or tying in the operations with a results timeline, which can show when orders were received or sent. Finally, the user in the above mode can modify the process by changing the staffing levels at the various units or the processing time required by the agents.

Given these capabilities, the above mode of play could help train users to understand the structure of the overall process, as well as to examine the ripple effects of modifications made to staffing parameters, allocation of resources, and changes in policy and procedures.

Within mode. During the simulation, the computer agents in the various units receive, process, and generate orders in the form of text files in the computer memory. Because these text files are also readable by the human user, the user can perform these very same activities. Thus, in the within mode, the user can play as an agent in any one of the units, using any type of text editor to read and write files (simple but notional orders).

This mode can be played in a number of different ways, depending on the types of lessons to be learned. The user may play the simulation either with or without the full process picture (which shows how the activities flow toward him and away from him). The user may be allowed to batch orders. For example, if he receives two orders for a certain type of supply going to the same final station, he could combine them into a single order. The user could be made to respond to priorities, reshuffling the flow of orders handled to respond to their respective priorities. The user could be made to consider delays in the process, so that he can improve overall performance by holding off the processing of a particular order until a parallel activity track can "catch up." Or the user may be forced to track down and correct mistakes that have been generated elsewhere in the process.

The overall goal of this type of play would be to reinforce the conception of the process structure and operation by actually putting the user "in the action."

Networked mode. Because the simulation generates and uses standard text files during operations, standard file sharing protocols (on PC, Macintosh, or Unix networks) could be used to have a networked game with multiple users playing the same scenario. In this networked mode, each player would play a unit, monitoring a mailbox (file folder) and receiving, processing, and sending orders. These players could run the simulation in a single room or at separate locations distributed over a wide area network (WAN).

The same sorts of permutations possible in the within mode could be used in the networked mode. Thus, for example, individual users could play the game with or without the full process picture, and the game could be played with or without random process errors. In addition, the networked structure would allow play with or without full communications (e.g., via telephone) with other players.

This sort of play could further reinforce the process lessons learned from individual play in either the above or within modes. In addition, this mode could support the conception of the process as a networked structure and demonstrate how communications could be used to improve the operations of the process.

Contingency Operation Class IX Distribution Network Microworld Model Prototype

This prototype microworld simulation is generally patterned after U.S. Army Europe's (USAREUR's) Class IX distribution network used to support OJE during the first half of 1996. Part of our team actually "walked the process," starting at Ramstein Air Force Base and following the truck route from Germany to Task Force Eagle in Bosnia. The composite network diagram depicted in Figure 4.5 is not intended to be exact; rather, it is illustrative of the information provided during interviews with each staff responsible for a particular part of the network. The Army's relatively new Battlefield Distribution System is partially represented by the USAREUR's hub-and-spoke concept, with the primary theater hub at the Kaiserslautern Industrial Center to sort breakbulk cargo arriving into Germany or France by air. The hub had to manage the flow to peacetime garrison locations for units remaining in Germany and to the deployed elements in Hungary at the Intermediate Support Base (ISB) and in Bosnia.

We created a prototype microworld model of the USAREUR Class IX Distribution Network used to support Operation Joint Endeavor to illustrate the "ripple effects" of decisions made elsewhere, providing a projection of how decisions made elsewhere would affect a unit's mission at a particular node. We used the prototype microworld model to examine different transportation policies within a scenario similar to OJE. One design represented an alternative

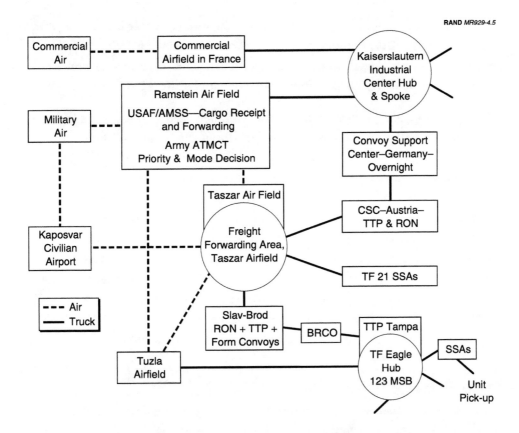

Figure 4.5—Composite Network Diagram of USAREUR's Class IX Distribution Network Used to Support OJE in June 1996

to the OJE design. In this alternative design, the model moved cargo using trailer transfer operations to determine how long it would take for supplies to arrive in Hungary after departure from Kaiserslautern. From the hub to the ISB is less than a three-day trip, if measured by driving time alone, and trailer transfer operations significantly reduced the time from Kaiserslautern to the ISB. In exploring this policy during a training event, the staffs need to have information about the operational situation, such as that in Bosnia; during the initial stages of OJE, TF Eagle support units could not handle the large flow of cargo, so there was no urgency in getting supplies delivered by truck any faster. Critically needed supplies were flown directly from Ramstein Air Force Base to either Taszar, Hungary, or Tuzla, Bosnia.

Another operational consideration for staffs involved in microworld process training is the allocation of assets between major organizations. For example, at the start of OJE, the 37th Transportation Group "hub" was not staffed or equipped to provide service both within Germany and to the deployed forces in Hungary and Bosnia. The German army provided trucks and drivers to operate on "spokes" within Germany to augment the U.S. Army capabilities as other U.S. assets were allocated to OJE. Corps assets were also reallocated during OJE. Some V Corps transportation assets were assigned for operational command and control to TF Eagle and stationed in Bosnia, rather than being under the ISB control in Hungary.

Microworld models could be used to expose staffs to such processes in a training environment so that they might consider the ripple effects caused by certain policies. Staffs would learn how to work out processes for actual deployments to attenuate such ripple effects rather than react to them in an ad hoc fashion. Ripple effects will not disappear, but developing a microworld process will help staffs better prepare for those situations requiring an ad hoc response. Such a learning environment would help them understand how to design, examine, and redesign their processes to reduce the number of ad hoc situations in advance. This is not a new idea: It is the purpose of planning. What is unique is that the microworld simulation provides the capability to rapidly evaluate a large number of different cases, allow changes to the process, and provide visibility and measurement of how changes in each segment affect the whole process.[2]

Our diagram does not illustrate the use of radio frequency tags to track shipments throughout the network. However, we are experimenting with methods to incorporate this real-time data flow into a microworld simulation environment. The important aspect of these experiments is to create a situation in which staffs can learn how to deal with information technology—how to incorporate the data flow into their processes and transform the data into information for monitoring the network and for helping to shape plans and allocation decisions for other phases of a contingency operation.

Pilot Studies of the Prototype Microworld-Model Curriculum

As mentioned earlier, we conducted pilot studies of the curriculum centered on the microworld prototypes we developed. Below, we discuss the background for the pilot studies, including goals, approach, procedures, and measures; then, we discuss what we learned in conducting them.

---

[2]Our illustrative diagram alone does not provide the full picture of operational aspects. The 37th Transportation Group's long-haul trucks went primarily from Germany to the ISB in Hungary. The trip from Germany to the ISB in Hungary normally took at least three days because of the way that the transportation assets were used. A truck would leave Kaiserslautern, remain overnight near Regensburg, Germany, at a convoy support center, then spend a second night near Vienna, Austria, at another convoy support center, before embarking on the third day of the trip. The Vienna convoy support center was used both as an overnight stop for 37th Transportation Group trucks going directly to Hungary and as a trailer transfer point between 37th Transportation Group and one of the transportation battalions assigned to the ISB.

Pilot Study Background

Goals. The primary goals for the pilot studies were to first observe and document how various new methods or training activities (described in this document) enhance the current training environment for the 310th TAACOM, the Force XXI experimental unit for CSS above the corps level. This included investigating the benefits of activities such as mapping the "processes" managed by a TSC in a small-group process format and using microworlds. We also determined, through observations and interviews, what aspects of the microworld models were viewed as most important for training.

Evaluating the effectiveness of the new methods for teaching CSS management knowledge and skills was at the heart of the pilot studies. As such, we developed appropriate evaluations of training content, including both traditional and more "hands-on," or dynamic, assessments of how well a staff was able to understand and manage a complex CSS system.

Important to understanding the cost-effectiveness of any new curriculum or change in training methods is understanding the implications for resources: What changes will be required of training developers' time, expertise, and tools? Will training implementors require more or less time? How will time and cost burdens change or be shifted from institutions to units, or vice versa? Our pilot studies provided the foundations for beginning to address such important questions.

Approach. The approach we used was first to refine the microworld models through use by small groups of subject-matter experts (SMEs) on TSC doctrine and operations. A variety of activities are planned for player groups, based on both process mapping exercises and experiences with the dynamic microworld models.

Procedures. As noted earlier, large simulations take weeks to months to set up and play, and then they only represent a small portion of the unit's mission. We demonstrated in the pilot study an approach to conducting training in segments that cover the breadth of a unit's mission within a training program that addresses unit personnel turbulence as well as completeness.

The goal we set was to conduct a more comprehensive course of training that included the breadth of missions over a longer simulated time horizon during the same two-week period (ten 12-hour days) in which the current system trains basically one mission for about three days of a contingency. Currently, the unit gets at best about 60 hours of functional training (five 12-hour days), with the remaining 60 hours focused on how to operate the simulation and understand artificial aspects of the model and the exercise. For example, we proposed packaging a training event in a 4-hour segment that would simulate 30–90 days of CSS operations rather than three days. We also proposed that this microworld-based curriculum cover multiple 4-hour segments during a reserve unit's two-week annual training; thus, during the 120 hours of training time available (ten 12-hour days), we could achieve significantly more than the 20 hours of simulation time that many units have been achieving just to represent three days of an operation.

Measures. The training events needed to enhance the quality of learning. Further, the tools used in the training event needed to be adaptable to a variety of units and situations to increase the opportunity of staffs to learn in an environment of frequent personnel turnover.

Conducting the Pilot Studies

Here, we discuss the pilot studies themselves and then turn to what we learned in conducting them.

Pilot studies and participants. We conducted pilot studies with both individual staff members and small to large groups. With individual staff members, we discussed the curriculum, demonstrated the prototype model, and conducted interviews to obtain suggestions for improving the representation of CSS functions in both the curriculum and the prototype model. We organized the sessions within the context of the units' preparation for participating in U.S. Army Japan's bilateral exercise Yama Sakura XXXIII, held in January 1998.

The earliest pilot studies with the small and large groups were conducted at the 310th TAACOM during two weekend drills in July and August 1997. These studies consisted of meeting in a large group, which varied from 35 to 50 people, and then breaking down into smaller groups of 10 to 15 people. These sessions were conducted during a four-hour period; however, because of other administrative functions, breaks, etc., actual training consisted of about two and a half hours. Both group sessions comprised members from the 310th TAACOM headquarters general staff and support operations staff; the second group session included members from the 4th Movements Center in addition to the 310th TAACOM staff.

In December 1997, we conducted a four-hour session with two Army officers who were assigned as Research Fellows by the Army to RAND'S Arroyo Center. This session was focused on the content and relevance of the training materials to

Army staffs; the judgments of these officers were useful in refining both the training content and our data collection and assessments instruments.

In January 1998, we conducted a session with the 311th Corps Support Command. This session comprised fifteen staff members, and the intention was to finalize our approach and training materials for the more comprehensive pilot study conducted in May 1998 during an actual TSC exercise.

Since the TSC is an emerging Force XXI organization, we were not sure at the outset of the pilot studies how much to focus on its structure and missions. As we learned from the pilot sessions, we continued to refine the content of our final prototype training session. We wanted the focus to be on learning systems–dynamic thinking rather than organization structure. Consequently, we reduced the amount of time spent on TSC mission and organizational structure and found that this had little impact on learning. We also varied the amount of time devoted to the section on measurement and diagnosis, observing that we could present a small measurement and diagnosis section as a basic enabling skill and then go on to illustrate measurement and diagnosis more comprehensively in the model portion of the curriculum, which focused on a dynamic situation.

What we learned. There were several significant outcomes from the perspective of the study participants. Both the participant discussions and the written assessment questionnaires indicated trends in the right direction:

- Microworld models do help with more complex planning and coordinating tasks, which is where we designed them to help.

- The four-hour training sessions provide a good learning environment. In fact, about half the respondents in the 310th TSC (P) studies indicated a desire for a series of small-group sessions focused on their staff functions.

- Each segment of training—briefing and discussion of TSC organization, paper analysis of scenario, and microworld model analysis of scenario— showed effects in the appropriate area. The organizational and exercise-specific training aspects showed less improvement over current training, while the microworld model aspects showed more.

The pilot studies provided us with sufficient information and confidence in the training approach to enable us to continue planning larger-scale experiments, which we conducted with a large group during 310th TSC (P) Early Entry Module-Exercise 1998 (EEM-EX 98) in May 1998.[3] Specifically, they helped us to

---

[3] Again, the detailed methodology and results of this successful larger demonstration will be reported separately in another document.

gauge our materials, questionnaires, and training schedule. For example, we observed that our goal of conducting a compact four-hour session could be implemented from the perspective of training content, but that its feasibility depends on other factors. One of the limiting factors is the background preparation of participants. Since the TSC organizational structure, doctrine, and SOPs have not been solidified by the Army, we had to spend more time than desired instructing the participants on the emerging doctrine about the role of the TSC. This detracted from the other curriculum elements, especially the time available to conduct multiple iterations of the microworld simulations.

Judging in advance that this would occur, we prepared a CSS Battle Book for the four-hour sessions, which was provided to each participant. This Battle Book included the background on the TSC, on the Yama Sakura exercise, and on our research efforts to date. We found this a necessary step before proceeding to discuss the mission essential tasks, which we intended to teach using a microworld model as a tool to understand dynamic complexity.

5. Conclusions and Implications

Below, we briefly summarize our key conclusions and present some implications of the research for other organizations.

Conclusions

Force XXI, the Army's ongoing effort to reengineer itself, has placed a new emphasis on the dynamic complexity of the battlefield and on the CSS aspects of information operations for the digital battlefield. Many of the Force XXI technologies being developed as CSS enablers are within the realm of the possible today and could, if sufficient funding were allocated to their final development and acquisition, be fielded over the next few years in advance of other Force XXI initiatives.

However, the training to change the mindset of current CSS staffs to address the concepts of Force XXI is absent. This absence is reflected in the current "organizational/doctrinal" approach to training, which is evident in both the training materials and simulations being used. Specifically, the Army's traditional approach to writing staff training doctrinal manuals centers on a "stovepipe" view—how to train within a given organization. Since the organizational concepts for Force XXI are still in flux, however, the traditional method for developing training materials does not seem appropriate. In addition, the "checklist" approach of current MTPs is useful to center the discussion, but that approach does not challenge the staff to be more proactive in CSS planning.

The simulations being used are also inadequate to address the changes being effected by Force XXI. The Army has been experimenting primarily with the brigade-level organization and operations and emphasizing the combat operations aspect, not the CSS aspects. And even in the Army Warfighting Experiments (AWEs) conducted to date that have emphasized CSS operations at EAC, the focus of the exercises is on organizational/doctrinal aspects and not on process; consequently, they do not capture the robustness required to support the Army's planned "full-dimension operations."

Army training should emphasize the dynamic complexity of CSS aspects that support different phases of combat operations, which means that the Army needs a new training strategy—one that addresses structure, content, and methods and that shifts away from the organizational level and toward the processes that are the core of CSS operations. The Army is still refining the organizational structure, authority, and enabling technologies that will affect the efficiency and effectiveness of performing core functions. However, the basic CSS process elements and necessary management skills are sufficiently well understood at this time to enable us to develop a training strategy.

Specifically, using dynamic modeling tools similar to the prototypes we have developed and adopting a new curriculum strategy would provide a training environment in which staffs can experiment with different policies and business practices. Such tools would also be beneficial in AWEs for CSS organizations above corps because they would enable staffs to examine the impact of decisions in an end-to-end systems approach. Concurrently, the microworld model tools available in the commercial sector could be used in the new curriculum to train Force XXI CSS operational concepts and to reflect on the CSS processes necessary to implement those concepts in a contingency.

Our proposed approach is also consistent with the need to develop training programs in a significantly shorter time than before. Over the next year, we will be conducting more extensive experiments to demonstrate the effectiveness of our proposed approach and to develop a more detailed framework that highlights the impact of this approach in deriving more and better training in an environment of keen competition for increasingly diminishing resources.

Implications

While our focus here has been on the Army's CSS community, our research has applicability to any organization seeking to train staff interactions in policy development and measurement under conditions of uncertainty and in a distributed decisionmaking environment. Our prototype staff planning model goes beyond the level of individuals running a simulation in examining assembly line/manufacturing operations and beyond the business process reengineering construct that is part of some commercial software packages today.

In terms of the former, Womack and Jones cite in their 1996 book *Lean Thinking* the example of a Japanese management consulting guru coming into a company's manufacturing operation and physically moving machines around to demonstrate how a new process ought to work. While this may work in an assembly line operation, it would be highly unlikely in a setting akin to military theater operation. Instead, businesses use focus groups to orient on a particular policy application within their area of expertise, to see how their policies might play out. In terms of the latter, Senge et al. in *The Fifth Discipline Field Book* discuss the use of "flight simulators by small groups" to examine alternative policies.

Our proposed approach goes beyond such small groups and enables staff training to be conducted in a distributed fashion. The use of system dynamic models helps staff further understand the complexity of unfolding operations as they may impact on other interest groups in other organizations. Specifically, we have embedded the Extend model in a "gaming" environment that enables individuals to play from "within, above, and between" modes.

In doing this, we use a "glass box" approach: The players can actually see the underlying rules and change the rules as they change assumptions about the environment in which they can expect to be operating. In most current simulations, players usually can change variables readily, but *not* the underlying rule structure. If they want to change the underlying rules, they must normally be assisted by professional simulation/computer experts. The value of "flight simulators" in the "glass box" construct is not to understand the "right outcomes," but rather to understand how the players can design policies and test their robustness in terms of both total system outcomes and individual node performance.

Since many private-sector organizations operate over distant areas, as Army organizations do, bringing people together in a large game at a central location is both time consuming and extremely resource-intensive. We have demonstrated a curriculum and a prototype model that can be implemented under distributed conditions. We have demonstrated in pilot studies that staff training can be more effective by using our recommended approach. Instead of using a general, large-scale simulation once every two years, a unit can use a series of smaller-scale simulations targeted to its needs more frequently throughout a training year and in an annual exercise. The sample training audiences found our recommended approach and prototype training models acceptable and valuable in such an environment.

Bibliography

Hardcopy

Carr, David K., and Henry J. Johansson, *Best Practices in Reengineering*, New York: McGraw-Hill, 1995.

Draft Concept for Support Command and Control at Echelons Above Corps, Fort Lee, VA: U.S. Army Combined Arms Support Command, January 31, 1996.

Glennan, T. K., S. J. Bodilly, M. W. Lewis, D. J. McArthur, and J. Moini, *Background Materials on Agency Learning Technology Programs in DoD, NSF, and DoED*, Santa Monica, CA: RAND, PM-800-OSTP, 1998

Gray, W. D., D. B. Pliske, and J. Psotka, *Smart Technologies for Training: Promise and Current Status*, Alexandria, VA: Army Research Institute for the Behavioral and Social Sciences, 1985.

Krause, Larry, Simulation Dynamics, Inc. (SDI), responding to RAND inquiry about use of Extend in training environments, letter dated August 11, 1998.

McGrath, John J., and Michael D. Krause, *Theater Logistics and the Gulf War*, Alexandria, VA: U.S. Army Materiel Command, 1994.

McKenna, Regis, *Real Time: Preparing for the Age of the Never Satisfied Customer*, Boston: Harvard Business School Press, 1997.

Senge, Peter M., Charlotte Roberts, Richard B. Ross, Bryan J. Smith, and Art Kleiner, *The Fifth Discipline Fieldbook: Strategies and Tools for Building a Learning Organization*, New York: Doubleday, 1994.

Womack, James P., and Daniel T. Jones, *Lean Thinking: Banish Waste and Create Wealth in Your Corporation*, New York: Simon and Shuster, 1996.

Web Sites[1]

Gensym Corporation's primary Web page is at *http://www.gensym.com.* Its customer "success stories" can be found directly on that page.

High Performance Systems, Inc.'s primary Web page is at *http://www.hps-inc.com.* Selecting "About HPS" leads to "Learning Environment Products." Selecting one

---

[1]These sites list products, customer lists, and in some cases "customer success stories" related to the use of the sites product or services. We do not endorse the products or services, but report them here as part of our research effort.

of the products, e.g., "Building Service, Driving Profits: RGP Financial Services" leads to "Harvard Business School Publishing" *(www.hbsp.com/frames/groups/ cases/new/new_products.html)*.

Imagine That, Inc.'s primary page is *http://www.imaginethatinc.com*. The page links to Extend product information and to a "Training" page, which has links to several companies using the EXTEND application; each company lists its "success stories." We looked at Simulation Dynamics, Inc. (listed below separately) and Computer Aided Process Improvement *(http://www.capi.net)*, which lists the U.S. Postal Service and other customer case studies.

A game called Head Trader™ can be found from NASDAQ's primary Web page *(www.nasdaq.com)*: Pick "Site Map," then pick "NASD Related Web Sites," and then click on "Academic Research Web Site." This will bring up a screen listing Head Trader™ Game; click on that icon and it will switch to the game main entry menu (as of December 10, 1998, that Web page menu is at *http://www.nasd.com/ HeadTrader/head_trader.htm*).

Powersim Corporation's primary Web site is *http://www.powersim.com*. Customer "Success Stories" were downloaded from Web pages at *http://www.powersim.com/html/f_success_nexus.htm*, *http://www.powersim.com/html/f_success_ford.htm*, and *http://www.powersim.com/html/f_success_ford.htm*.

PROMODEL Corporation's primary Web site is *http://www.promodel.com*. A "Click here for help" icon leads to *http://www.promodel.com/corpguide.html*; then, choose the particular model in which you are interested. For example, picking "ProcessModel" leads to the page *http://www.processmodel.com*. The "Customers" link leads to *http://www.processmodel.com/customer/stories.html*. From there, select the particular company success story of interest.

Sandbox Entertainment Corporation's primary Web site is *http://www.sandbox.net*; further information was downloaded from its Web page *http://www.sandbox.net/ sandbox/pub-doc/main.html* and *http://www.sandbox.net/finalbell/*.

Simulation Dynamics Inc.'s site is at *http://www.simulationdynamics.com*.

The Society for Organizational Learning's site is at *http://learning.mit.edu/ index.html*. Research papers, proposals, and ongoing research can be found at *http://learning.mit.edu/com/peo/ediehl.html* and at *http://learning.mit.edu/ res/wp/ pubs.html*.

Discussions

Krause, Larry, Simulation Dynamics, Inc. (SDI), and John Bondanella, RAND, telephone discussion, August 11, 1998, concerning RAND inquiry about the use of Extend in training environments.

Shapiro, Roy, Harvard Business School, and John Bondanella, RAND, telephone and e-mail discussions during December 1996 through February 1997 about the pedagogical aspects of the Extend software model tools. Dr. Shapiro also provided his curriculum illustrating the use of the Extend model in his classes about manufacturing processes.

Stanton, Matt, Sandbox Entertainment Corporation, and John Bondanella, RAND, telephone discussion, August 10, 1998, about RAND inquiry about the use of Final Bell and other simulations in military and business training.

Seminars

Human Resources Round Table program, University of California, June 4, 1996. Ms. Emily Larson attended as our project's RAND representative. Speakers included James W. Candler, Managing Director, Personnel Information Systems, Federal Express; and Alexandra J. Rand, Founder and President, Internal & External Communications, Inc (IEC). Ms. Rand discussed LEAP (Leadership Evaluation & Awareness Process), an interactive multimedia program developed by IEC.